D0893916

ANNEXE DE LA BIBLIOTHEQUE
Université d'Ottawa
BIBLIOTHECA
LIBRARY ANNEX

QUATERNARY LANDSCAPES IN IOWA

QUATERNARY

IOWA STATE UNIVERSITY PRESS

LANDSCAPES IN IOWA

Robert V. Ruhe

AMES, IOWA, U.S.A.

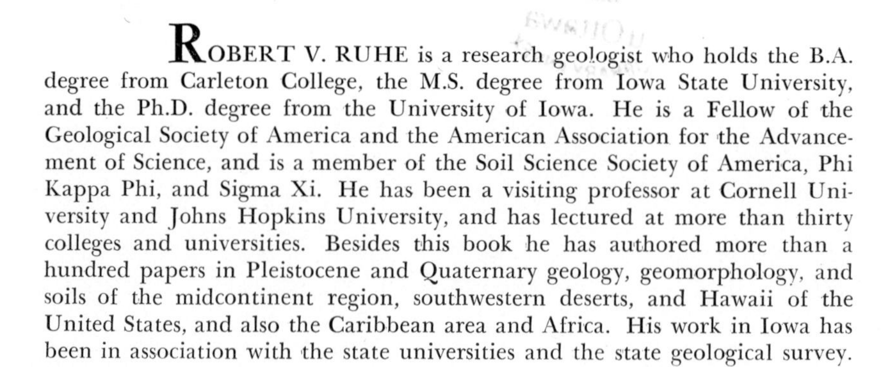

ROBERT V. RUHE is a research geologist who holds the B.A. degree from Carleton College, the M.S. degree from Iowa State University, and the Ph.D. degree from the University of Iowa. He is a Fellow of the Geological Society of America and the American Association for the Advancement of Science, and is a member of the Soil Science Society of America, Phi Kappa Phi, and Sigma Xi. He has been a visiting professor at Cornell University and Johns Hopkins University, and has lectured at more than thirty colleges and universities. Besides this book he has authored more than a hundred papers in Pleistocene and Quaternary geology, geomorphology, and soils of the midcontinent region, southwestern deserts, and Hawaii of the United States, and also the Caribbean area and Africa. His work in Iowa has been in association with the state universities and the state geological survey.

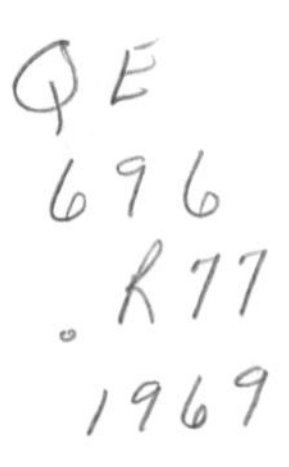

Composed and printed by
The Iowa State University Press

First edition, 1969

Standard Book Number: 8138–1321–2
Library of Congress Catalog Card Number: 69–14312

TO BARBARA MILLS RUHE

Who has not only shared but endured the Quaternary from Kakitumba to Ketchikan and beyond

PREFACE

THIS BOOK HAS BEEN BROUGHT TOgether for many reasons, of which the foremost is a summarization of the most pertinent aspects of the Quaternary of Iowa. The episodes of glaciation during the Pleistocene, and the features and events of the Recent or since the last glaciation, comprise the Quaternary in this state. More than two decades have passed since the last summary was written, and that, in turn, was a threefold work which spanned another two decades (Kay and Apfel, 1929; Kay and Miller, 1941; Kay and Graham, 1943). An objective herein is to bring the knowledge up to date as understood in the current state of the art.

Previous studies in Iowa have mainly emphasized the stratigraphy and geographic distribution of Pleistocene deposits and their associated glacial or glacial-related origins. Little attempt was made to introduce geomorphology into the overall system. This current writing will be reoriented toward an explanation of the landscapes of Iowa as formed in the Pleistocene and Recent. Since integral parts of the landscapes are soils, the problems will be tackled from the points of view of geomorphology and pedology. These will be put into the Pleistocene and Recent. Stratigraphy, geography, and many other things will not be ignored but will be used as needed.

One of the main objectives is to explain the landscapes on which we live and work.

During the past 15 years coordinated studies in geomorphology and soils have been directed toward trying to explain what effect the processes of evolution of the landscapes have on the formation of soils on that landscape. These studies have been carried out in many states, but the prototype was done in Iowa (Ruhe, Daniels, and Cady, 1967). Many principles have been formulated through the years, but they are scattered throughout the technical literature, and a private-eye would be needed to find them. Another objective, then, is to bring these things together in one place.

Recently a new school of geomorphic thought under the calling of "dynamic equilibrium" purports the time-independence of things concerning the landscape. An opposite view will be presented to show the importance of time or time-dependence of processes giving rise to the landscapes and soils of Iowa. They can be better understood within a time framework.

Fortunately, radiocarbon dating spans a part of the late Quaternary and all of the Recent and permits the establishment of an absolute calendar to which starts, durations, and stops of things can be related. These starts and stops of things, by the way, have been called "cycles," but this is a bad word today in geomorphology. Anyway, an additional objective will be to bring together all of the radiocarbon chronology of Iowa that is pertinent geologically and pedologically and weave the history.

In order to carry out the objectives, a basic principle in geology will be applied, that of uniformitarianism, which states that "the present is the key to the past." Most people like to return to James Hutton (1726–1797) and John Playfair (1748–1819) for this idea, but why go back beyond Samuel Clemens of Hannibal who wrote:

> Now, if I want to be one of those ponderous scientific people, and "let on" to prove what had occurred in the remote past by what has occurred in a given time in the recent

> past, or what will occur in the far future by what has occurred in late years, what an opportunity is here! . . .
>
> In the space of one hundred and seventy-six years the Lower Mississippi has shortened itself two hundred and forty-two miles. This is an average of a trifle over one mile and a third per year. Therefore, any calm person who is not blind or idiotic can see that in the old Öolitic Silurian Period, just a million years ago next November, the Lower Mississippi River was upward of one million three hundred thousand miles long, and stuck out over the Gulf of Mexico like a fishing rod. And by the same token any person can see that seven hundred and forty-two years from now the Lower Mississippi will be only a mile and three quarters long, and Cairo and New Orleans will have joined their streets together, and be plodding comfortably along under a single mayor and a mutual board of aldermen. There is something fascinating about science. One gets such wholesome returns of conjecture out of such a trifling investment of fact.[1]

Mark Twain well, for we shall have to travel far afield from Iowa to reason by analogy.

Distinct effort has been made in this state to build a pertinent radiocarbon chronology. Many colleagues have been involved. Among them are R. I. Dideriksen, W. P. Dietz, T. E. Fenton, G. F. Hall, J. D. Highland, F. F. Riecken, E. C. A. Runge, W. H. Scholtes, and P. H. Walker. They are noted in the Radiocarbon Catalog, Chapter 6, and thanks to them. Some of the dating was accomplished under National Science Foundation Grant GP-2610.

Pollen analyses referred to were done by Grace S. Brush and L. H. Durkee. All wood samples reported were identified by D. W. Bensend. Thanks to them also.

The quotations from *Life As I Find It* are used by permission of Charles Neider.

When the professionals have turned in for the weekend or a long holiday, my special field crew members, Barbara, Robin, Jon, Deborah, and Bill, have dug, pounded seismic hammers, sledged electrodes, and even carved their ways

[1] Mark Twain, *Life on the Mississippi,* Hill and Wang, Inc., New York.

through a wall of permafrost with a Bowie knife. Someone ran off with the pick and shovel. Thanks to them.

A special thanks to Mildred Griffin whose practiced eye and hand have handled my manuscripts for more years than she would care to admit.

R. V. Ruhe

CONTENTS

Part 1: THE BRIEFING

Part 2: MEMORANDA

PART 1

THE BRIEFING

THE LAYER IMMEDIATELY UNDER OUR layer is the fourth or "quaternary"; under that is the third, or tertiary, etc. Each of these layers had its peculiar animal and vegetable life, and when each layer's mission was done, it and its animals and vegetables ceased from their labors and were forever buried under the new layer, with its new-shaped and new-fangled animals and vegetables. So far, so good. Now the geologists . . . state that our own layer has been ten thousand years forming. The geologists . . . also claim that our layer has been four hundred thousand years forming. Other geologists, just as reliable, maintain that our layer has been from one to two million years forming. Thus we have a concise and satisfactory idea of how long our layer has been growing and accumulating.

MARK TWAIN
"A Brace of Brief Lectures on Science"
Life As I Find It

CHAPTER 1

BACKGROUND

IF ONE IS CURIOUS ABOUT THE LANDscapes on which he lives, works, or visits, he may understand them within a framework of five questions that may be asked about them: (1) What and (2) where are they? (3) How, (4) why, and (5) when did they form? Answers to the first two questions aid in defining the landscape in space. Answers to the second two questions will give some understanding of the processes and causes of formation. Answer to the fifth question aids in defining a part of the earth's surface in time and will also develop a historic system within which various parts of the landscape may be fitted.

Answers to the basic questions are derived through applications of various fields of earth science. One field, *geomorphology* (from the Greek, *ge, morphos,* and *logos,* meaning earth, form, and science, respectively), embraces all of the questions. However, other sciences must be called on to solve the problems. The landforms that together comprise a landscape have external three-dimensional form. *Geometry* becomes involved and is useful in the study of external form. The term *morphometric analysis* has been applied to such study. Soils are a rind on the external form of the landscape, and their study is *pedology* (from the Greek, *pedon* and *logos,* meaning soil and science).

A part of the landscape, such as a hill, has internal form. It is composed of minerals, sediments, and rocks; hence, *mineralogy, sedimentology,* and *petrology* require study. The sediments and rocks will have a specific arrangement within the hill, and their study introduces *stratigraphy.* They may contain shells, bones, or plant remains including microscopic pollen. So, *paleontology, paleobotany,* and *palynology* become parts of the study.

The materials comprising the landforms are subjected to *weathering,* that is, they are altered physically and chemically at or near the earth's surface. Study of this alteration from the geologic point of view is *geophysics* and *geochemistry,* but if restricted to the soil, is *soil physics* and *soil chemistry.*

Where time is investigated, *geochronology* is applied. This science has been stimulated greatly in the past several decades by the development of many methods of radioisotope measurement.

From the foregoing one may conclude that landscapes are complex features, and they are. However, one should not become frightened in attempting to study and understand them even though very high-sounding sciences are used in their study. The landscapes and the sciences are not that difficult.

We shall proceed to investigate kinds of landforms and emphasize the "when" about them. The other four questions that need asking will not be ignored. But first, a few basic things about landscapes must be introduced.

LANDFORM, LANDSCAPE, AND GEOMORPHIC SURFACE

Landforms are the features of the earth that together make up the land surface. They may be large features such as plain, plateau, and mountain, but may also be small features such as hill, valley, slope, and the like. Most landforms are products of erosion, but some of them are formed by deposition of sediments, by volcanic activity, or by movements within the crust of the earth.

According to a dictionary definition, *landscape* is the portion of the land surface that the eye can comprehend in a single view. More specifically, landscape is a collection of landforms.

Each definition is a relative thing. A hill has the components: summit, shoulder, slope, and footslope. From one perspective, each component may be considered as a landform. From another viewpoint, all of the components together comprise a geometric form and may be called a landform. A usable working rule is that landform should be handled as an individual or entity in a population. The investigator has the freedom to select a unit, and no difficulty arises if he delineates and defines the unit properly.

Landscape should be considered as a population. It is the sum of various kinds of landforms. In the previous example, the summit, shoulder, slope, and footslope may be grouped as a landscape that would be the valley slope. Another way of looking at the terrain would be to include repeated occurrences of the summed components and the result is a hill and valley landscape.

Such nonrigid definition has many advantages. Landform and landscape are independent of size and can be applied to very small or very large features of the earth's surface. Both terms are informal and should be used in that sense.

If formality is desired, *geomorphic surface* may be used, and it is a portion of the land surface that is specifically defined in space and time. It may occupy an appreciable part of a landscape and may include many landforms. It may also include many landscapes. The geomorphic surface must be a mappable feature. Its geographic limits and distribution in elevation must be delineable on aerial photographs or topographic maps. It has geometric dimensions which must be specified and which may be analyzed. It must be defined in association with other geomorphic surfaces in order to place it properly in its spatial and time sequence. It is associated with bedrock or sediments and may have bedrock or sediments associated with it. These associations must also be specifically defined in space and time. It is datable by relative or absolute means, and its dating must be specified. After definition of all of these relationships, the geomorphic surface is labeled, usually with a geographic name.

By interpreting the relationships of the geomorphic surface to other surfaces, rocks, or sediments, origin of the surface, whether erosional or constructional or both, may be determined. These are the broadest categories of generic classification. An erosion surface is formed by destructional processes active on a land surface, mainly running water. A constructional surface is built up by deposition in streams, ponds, lakes, and around ocean margins. It may be constructed by wind, glacier ice, tectonic disturbance which is the relative movement of one part of the earth's crust versus another, or by volcanic activity. The last mechanism is a spectacular event during which the birth and formation of landforms and landscapes may be observed.

For example, on the island of Hawaii the area between Pahoa and Cape Kumakahi, 23 miles south and east of Hilo, is very active volcanically. Near Kapoho Village numerous volcanic cones such as Halekamahina, Kapoho Crater, and Puu Kukae project above the surrounding rolling landscape of low relief (Fig. 1.1). In January, 1960, a rift or tear in the earth's crust occurred just north of Kapoho Village. Seven spatter cones formed from volcanic vents aligned along the rift (Fig. 1.2), and a new landscape was born. A main vent opened near the east end of the rift and built the main volcanic cone. A parasitic cone formed from a vent on the west side of the main cone (Fig. 1.2). Lava from the seven aligned spatter cones, and volcanic bombs and cinders from these cones as well as from the main and parasitic cone, destroyed the sugarcane plantation.

Molten lava breached the east rim of the main vent (Fig.

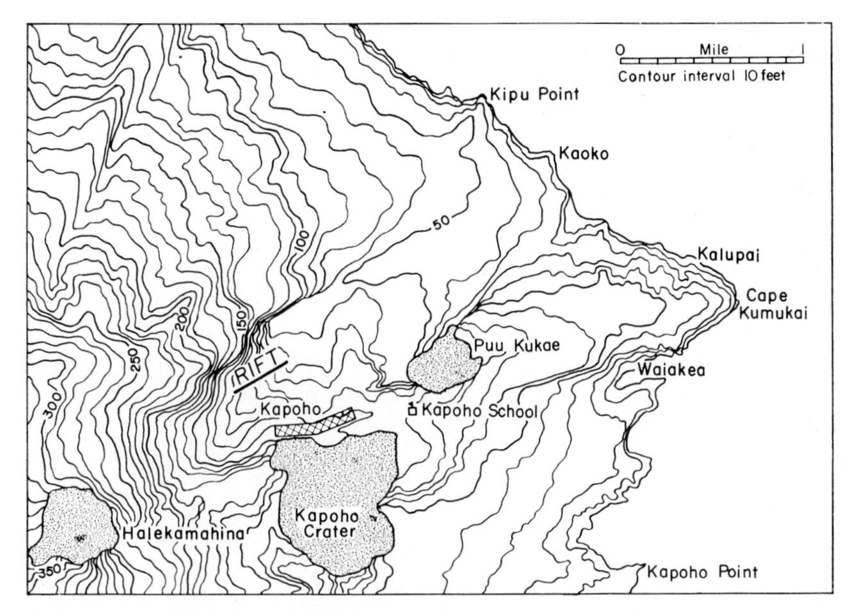

FIG. 1.1. Kapoho area on the island of Hawaii. Shaded parts are the volcanic cones of Halekamahina, Kapoho Crater, and Puu Kukae. Site of Kapoho Village in January, 1960, shown by cross hatching. Area just north of village was sugarcane plantation. Photographs of Figs. 1.2 and 1.3 are referred to this map.

1.2) and flowed northeastward downslope past Puu Kukae into the Pacific Ocean between Kalupai and Kaoko (Figs. 1.1, 1.2). Later, lava spilled over the shallow topographic divide between Kapoho Crater and Puu Kukae, destroyed Kapoho School, and descended to the sea (Fig. 1.1). Volcanic activity ended in February, 1960.

There is no question about the "when" of the construction of this landscape. It formed during January and February, 1960. The end result of the mountain building and lava surfaces, when viewed in June, 1965, some 5½ years later, is a barren, raw, rugged, and stark terrain (Fig. 1.3). The height of the new mountain and its size can be estimated by comparison to Paukea (Fig. 1.3). Elevation of the summit of the latter is 307 feet, and elevation at the base of the cone is

FIG. 1.2. (A) Seven spatter cones formed from volcanic vents aligned along rift. Main vent opened at east end of rift, forming major cone in right background. Parasitic cone is on west side of major cone. View from west end of rift looking eastward (Fig. 1.1). (B) Night photograph of fire fountains with main volcanic vent on the right and with parasitic cone on the left. View from Kapoho Village looking north (Fig. 1.1). (C) Fire fountain of main volcanic vent and lava flow from breached vent rim. View from Puu Kukae looking west (Fig. 1.1). (D) Terminus of lava flow spilling into Pacific Ocean between Kalupai and Kaoko (Fig. 1.1). View looking north. All photos several days after eruption, January, 1960.

FIG. 1.3. (A) Volcanic cone, age 5½ years. Destroyed Kapoho Village beyond newly constructed road. Note survival of some vegetation. View from Paukea, a small volcanic cone on north side of Kapoho Crater (Fig. 1.1). (B) Volcanic cone in right middleground with *aa* (blocky) and *pahoehoe* (ropy) lava flow surfaces in foreground. In left middleground is Paukea, a parasitic cone on the north side of Kapoho Crater. In center background is Halekamahina, an older cone, and in left center distant background is Kahaluaokahawale Crater, another older volcanic cone. View from west side of Puu Kukae, looking west. (C and D) Views sweeping right (north) of the volcanic cone and the lava constructional surface. (R. V. and Barbara Ruhe, June, 1965.)

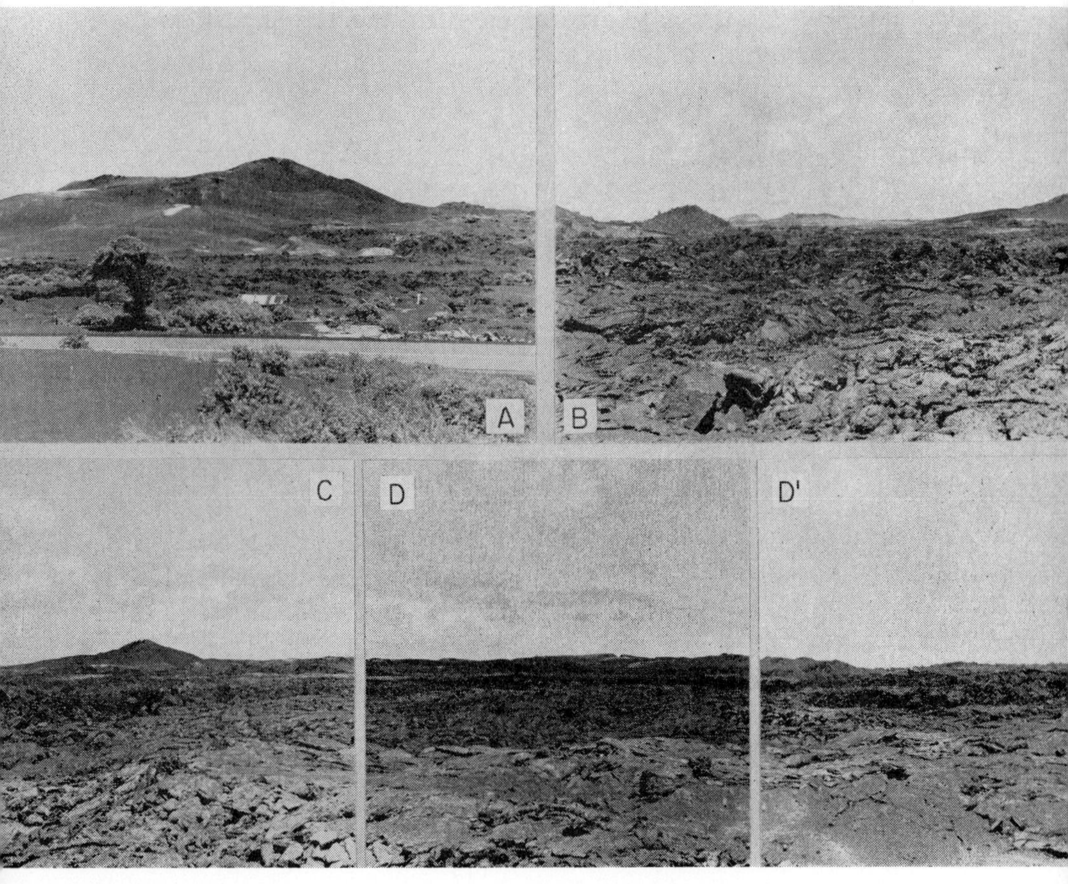

150 feet. The height is 157 feet. The summit of the new cone stands more than 200 feet above the general level of the surrounding lowland. The diameter of the base of Paukea is about 800 feet, so the diameter at the base of the new cone is about 1,800 feet. This sizable new landform was constructed in somewhat more than a month.

In 1960 the sugarcane was growing on soils formed by weathering processes that resulted in horizonated soil profiles with A and B horizons. In the same place several weeks later, a mountain and a rough, broken landscape of raw, unweathered basaltic rock was in place. Constructional processes can completely alter the landscape.

SOIL, SOIL PROFILE, AND WEATHERING PROFILE

The weathered rind of the land surface is an integral part of any landform or landscape. This association commonly is overlooked even in sophisticated research studies approached from the point of view of geomorphology or from the point of view of pedology.

Within the weathered rind, soils are at and beneath the land surface to variable depths. The base of the soil, however, may not be the base of the weathered rind. Commonly, zones of alteration continue downward to considerable depths beneath the land surface. Attention should not be focused solely on the vertical dimension of these features. Instead, they should be considered in three dimensions. On a hill, soils will occur up and down slope and along the contour, and altered zones may extend deeply beneath the surface into and through the interior of the hill.

In order to discuss the weathered rind of the earth, parts of it require definition and, at least, some brief description. However, a problem arises because there are no generally accepted definitions of these parts. Soil has almost as many definitions as the number of authorities who have written about it. If it is defined on the basis of its properties, soil is the outer layer of the earth's crust, usually unconsolidated, ranging in thickness from a mere film to more than 10 feet, which differs from the material beneath it, also usually unconsolidated, in color, structure, texture, physical constitution, chemical composition, biological characteristics, probably chemical process, and morphology (Marbut, 1951, p. 3). In relationship to plants, soil is the natural medium for the growth of plants on the surface of the earth, or is the natural body on the surface of the earth in which plants grow, and is composed of organic and mineral materials (*Soils and Men*, 1938). From a genetic viewpoint, soil is the collection of natural bodies occupying portions of the earth's surface that support plants and that have properties due to the integrated effect of climate and living matter, acting upon parent material, as conditioned by relief, over periods of time (*Soil Survey Manual*, 1951). In generalized equation form (Jenny, 1941), soil is the portion of the crust of the earth the properties of which vary with the soil-forming factors as:

$$s = (cl, o, r, p, t \ldots)$$

where s is soil; cl, climate; o, organisms; r, topography; p, parent material; and t, time.

All of these definitions leave something to be desired. Separation of soils from weathering zones would be difficult in using any one of them. Yet, traditionally pedologists work with soils, and geomorphologists work with weathering profiles. Infrequently their studies overlap. For convenience, pedogenesis or soil formation may be regarded as consisting of two groups of processes: (1) weathering that forms parent material, and (2) weathering that forms soil from parent material (Robinson, 1949). At most places, the two categories can only be arbitrarily separated.

In working with soil one must move from the generality, *soil*, to the specific, *a soil*. A parcel of soil has to be defined and delineated. The dimensions of a soil are fixed vertically by the thickness of the soil profile and laterally by practical limits of space required for observation. Vertically, a soil must extend from the surface downward into the parent material, and the unit must extend in two directions far enough laterally so that the field properties may be measured and sampling accomplished (Cline, 1949). A soil grades laterally to other soils with somewhat dissimilar properties. Consequently, a soil may be mapped and a boundary may be drawn around it.

The smallest volume that can be called a soil has recently been termed the *pedon* (Simonson and Gardiner, 1960). This may be conceived as a representative three-dimensional sample of a soil. The vertical dimension or a vertical slice of a pedon is the *soil profile*. This, in turn, may be thought of as the representative sample of the pedon (Fig. 1.4).

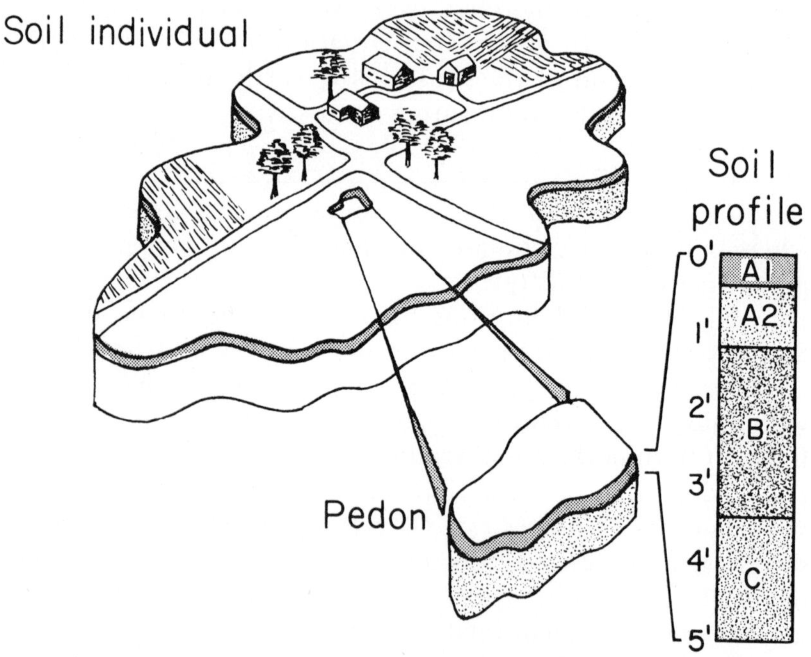

FIG. 1.4. Relationship of a *soil* or soil individual, *pedon,* and *soil profile.* (In part from Simonson, 1957.)

A soil profile may be quite simple or exceedingly complex. The profile is usually composed of soil horizons which are layers that approximately parallel the land surface and whose properties have been produced by soil-forming processes. Soil horizons are identified and differentiated on the basis of properties such as color, texture, structure, thickness, mineral and chemical composition, and the like. Certain master horizons are recognized and are broadly divided into organic and mineral horizons. Each of these, in turn, is divided into major kinds. The master horizons are symbolized alphabetically with upper-case letters and with Arabic numerals (Fig. 1.5).

The soil profile is the uppermost part of the *weathering profile* that is composed of zones of altered material extending downward to unaltered rock or sediment. The zones occur

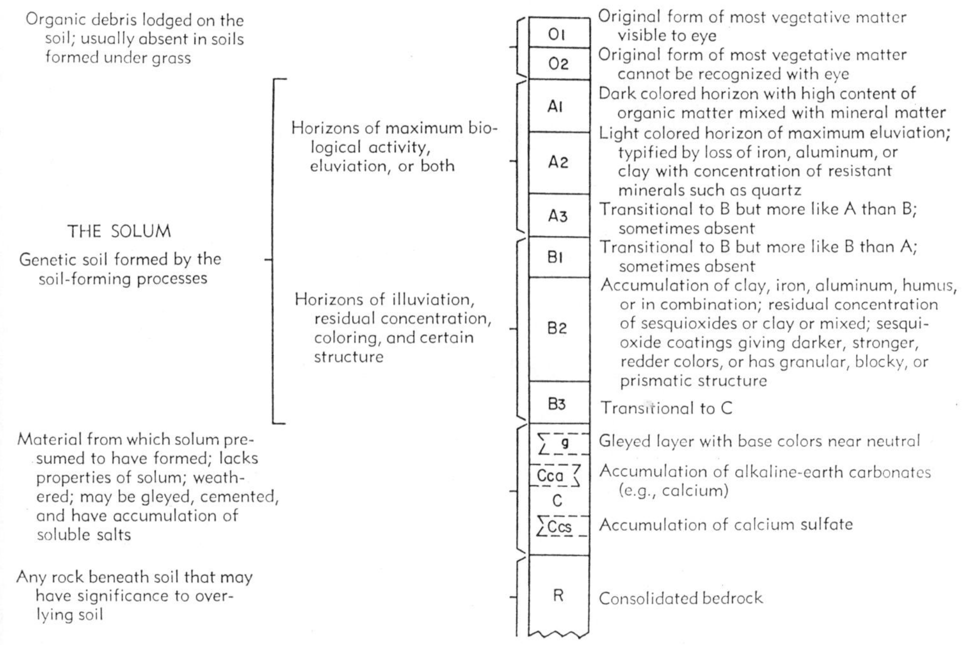

FIG. 1.5. Hypothetical soil profile that has all of the principal horizons. Not all of these horizons are present in any profile, but every profile has some of them. (Modified from *Soil Survey Manual,* 1951, and supp., 1962; and from Simonson, 1957.)

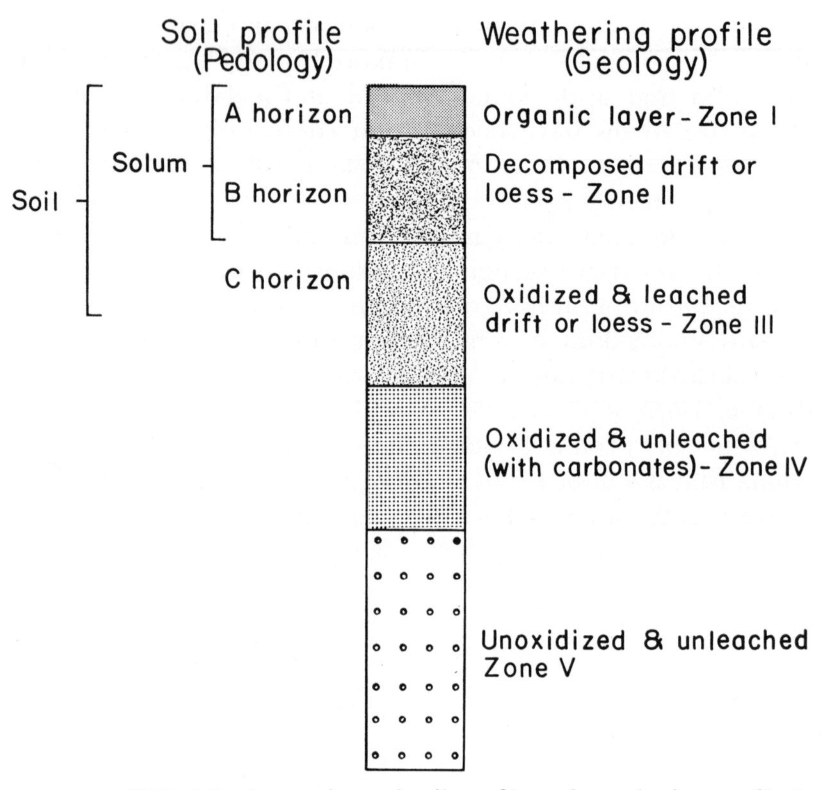

FIG. 1.6. Comparison of soil profile and weathering profile in surficial sediments in midwestern United States. The soil profile is the upper part of the weathering profile and may be composed of many different kinds of horizons in different combinations (cf. Fig. 1.5). The weathering profile is less complex and is usually much thicker than the soil profile.

in a vertical succession of layers that differ physically and chemically from each subjacent layer. The lowest zone is presumably the least altered and is the best representative of the original nature of the material. The relationship of the soil profile to the weathering profile may be illustrated diagrammatically (Fig. 1.6).

Weathering profiles are perhaps best known from Pleistocene geologic studies in the Upper Mississippi Valley Region. There, in glacial till, loess, and other sediments, distinctive zones of weathering are in layers beneath the land surface and have been known for decades as oxidized and leached, oxidized and unleached, and unoxidized and unleached zones. Oxidation refers to the nature of iron in the profile, and leaching refers to the presence or absence of calcium and magnesium carbonates in the profile.

More than 40 years ago the mineral and chemical natures of these zones were related to weathering alteration (Kay and Pearce, 1920). In the lowest zone limestone and dolomite are present (Table 1.1). In an overlying zone the more soluble carbonates have been leached with corresponding residual concentration of more resistant rock fragments such as granite and basalt. In the uppermost zone, even these more resistant fragments have weathered to form clay so that only the most resistant siliceous particles such as quartz and chert remain. Chemical properties verify the zoning. More soluble bases,

TABLE 1.1. WEATHERING PROFILE IN KANSAN TILL

Pebble content

Rock types	Weathering zones: Gumbotil*	Weathering zones: Oxidized and leached	Weathering zones: Oxidized and unleached
	(%)	(%)	(%)
Quartz	48.5	16.8	6.4
Chert	31.8	16.5	8.3
Quartzite	6.8	8.0	3.0
Granite	7.8	20.3	11.0
Basalt	2.9	24.5	27.0
Sandstone	0.5	1.0	1.0
Limestone	0.0	0.0	40.0
Others	1.0	11.7	3.8

Chemistry

Weathering zones	Composition: SiO_2	Fe_2O_3	Al_2O_3	CaO	MgO
	(%)	(%)	(%)	(%)	(%)
Gumbotil*	70.46	4.17	12.04	1.21	0.55
Oxidized and leached	71.84	4.62	10.86	1.29	0.72
Oxidized and unleached	66.56	4.40	11.13	4.48	0.79

Source: Kay and Pearce, 1920.
* Gumbotil is a paleosol and is the uppermost zone.

CaO and MgO, increase downward, indicating the leaching of carbonate in the upper parts of the profile. Distinct increases are in the unleached zone. Alumina is most abundant in the uppermost zone, showing the breakdown of silicates to form clay. Oxidation is illustrated by the relatively high and uniform iron-oxide composition to depth.

Many complications are involved in this very generalized statement about the weathering profile, in the relationship of weathering zones to landscape, and in the relationship of soil profile to weathering profile. These complexities will be introduced where pertinent and as needed.

All of the foregoing will form a basis for communication about landscapes. Because the major emphasis will be directed toward the "when" of things, a background is needed on principles and methods of dating or the determination of time or age.

DATING SEDIMENTS, LAND SURFACES, AND SOILS

Basically, there are only two principal ways to date earth features which are by relative and absolute means. In the first method one feature may be determined as younger, older, or the same age as another feature. This is mainly qualitative, but some scale of relative difference may be estimated such as "much younger or much older." However, how much younger or how much older cannot be determined without some kind of built-in chronometer in the system. Fortunately, the clock has been naturally built into many things in the form of radioisotopic elements. Radioactive decay of these elements and its measurement provide a means of absolute dating. A chronology can then be established by combining the relative and absolute methods of dating.

Relative Methods

One of the main interests herein is the dating of sediments. If these sediments have been deposited in an essentially horizontal attitude or with slight deviation from the horizontal, the principle of superposition applies. Younger beds are on top of older beds. Consequently, at any place the youngest bed will be at and immediately beneath the land surface, and successively older beds will be at greater depths. Distortion and deformation of the beds by earth movement may occur after deposition, leading to complex arrangements of beds. But where such disturbances have not occurred, the simple principle of superposition provides a relative dating scheme.

Even in areas of horizontal bedding, arrangements of beds in space may be more complicated than the type like layers of a cake. If a stream incises a valley and deposits sediment, this younger material may be a considerable distance in elevation *below* an older bed that is just beneath the land surface at the top of the adjacent hill. In the valley alluvium itself, a channel-fill deposit may be inset below the top of an older alluvial bed. These are only some examples of complications in the arrangement in space of essentially horizontal deposits. There are many others whose idiosyncracies will be introduced as needed.

Relative age relations of land surfaces to other land surfaces and to sediments are more complex. No difficulty will develop if the land surface is recognized in its true geometric nature which is a plane. This plane may vary from a simple two-dimensional rectilinear form to an extremely complicated curvate surface. In the foregoing example of the valley incised in horizontal beds, the valley hillslope deviates from the horizontal and descends obliquely across the beds. The hillslope surface is younger than the uppermost bed inside the hill. The hillslope also is younger than the land surface at the top of the uppermost bed, that is, the summit of the hill. The hillslope descends to the alluvial fill in the valley and must be the same age as the alluvium. In other words, hillslope erosion forming the hillslope surface provided the sediment that was deposited at the base of the hillslope. The relation can be further refined. Assuming that the valley alluvium is 20 feet thick, the maximum determinable age of the hillslope equals the age of the base of the alluvium. The minimum determinable age of the hillslope equals the age of the top of the alluvium. If a slope profile were drawn, one line, the cross section of the plane of hillslope, would bifurcate downward at the edge of the valley alluvium. From the bifurcation point, one line passes under the alluvium, and one line passes across the top of the alluvium. The record of the hillslope surface, a two-dimensional plane illustrated by a line, laterally becomes a three-dimensional body that is illustrated by a vertical plane.

Several important principles are involved in this example. A land surface is younger than the youngest deposit that it cuts, that is, the hillslope versus the uppermost bed in the hill. A land surface is also younger than the youngest other land surface that it cuts, that is, the hillslope versus the hill summit. A land surface is contemporaneous with alluvial deposits that lie on it. In regard to hillslopes specifically, the *principles of ascendancy and descendancy* are illustrated. A hillslope is the same age as the alluvial valley fill to which it descends, but is younger than the higher surface to which it ascends.

If erosion surfaces cross country, various relations may be used to determine the age of the surfaces (Trowbridge, 1921). (1) Any erosion surface is younger than the youngest material which it cuts. (2) It is younger than any structure that it bevels. (3) An erosion surface is younger than any material of which there are distinguishable fragments or fossils in alluvial deposits on the surface. (4) It is contemporaneous with alluvial deposits which lie on it, and (5) it is the same age as or older than other terrestrial deposits lying on it. (6) An erosion surface is older than valleys which have been cut below it, (7) is younger than materials forming erosion remnants above it, and (8) is older than deposits in valleys below it. (9) An erosion surface is younger than any adjacent erosion surface which stands at a higher level, and (10) is older than any lower adjacent erosion surface. These principles, upon analysis, become quite evident and need not be belabored further.

Absolute Methods

During the past several decades numerous methods in geochronology have been perfected or improved upon and are based on the radioactive decay of natural occurring elements. Some of these methods are uranium-lead, thorium-lead, potassium-argon, strontium-rubidium, radium-ionium, and, of course, the well-known radiocarbon method (Kohman and Saito, 1954). The main interest herein is in the last method.

All of these methods are based on the decay of a radioactive species which follows a rate law like that of a first-order chemical reaction. The rate of decay is proportional to the number of atoms (N) present at any time as:

$$-dN/dt = \lambda N$$

where λ is the radioactive decay constant. If N_o is the number of radioactive atoms present at time $t = 0$, then, by integration of the above equation:

$$N/N_o = \exp(-\lambda t)$$

indicating an exponential decay law. Instead of λ the quantity *half life* is commonly used. So, from the last equation

$$\tfrac{1}{2} = \exp(-\lambda\tau)$$

$$\tau = \frac{\ln 2}{\lambda} = \frac{0.693}{\lambda}$$

In regard to carbon (Libby, 1960, 1965), radioactive C^{14}

is produced by impingement of cosmic rays on the earth's atmosphere where the dominant gas N^{14} is converted to C^{14} by neutron bombardment as:

$$N^{14} + n = C^{14} + H^1$$

A radiocarbon isotope of mass 14 with a half life of 5,570 ± 30 years is produced. Radiocarbon ages are reported on this basis, although a half-life value of about 5,750 years is now believed to be more accurate. By combination with oxygen, a radioactive carbon dioxide forms and mixes rapidly with atmospheric carbon dioxide. Hence, all atmospheric CO_2 is made radioactive by cosmic radiation. Since plants use CO_2 in photosynthesis, all plants become radioactive. Animals eat plants and, in turn, become radioactive. Thus, all living things become radioactive.

While any organic matter is alive, it is in equilibrium with the cosmic radiation. All of the radiocarbon atoms which decay in an organism are replaced by C^{14} in photosynthesis or by eating plants. The specific activity is maintained at a level of about 14 disintegrations per minute per gram as expressed by:

$$C^{14} = \beta^- + N^{14+}$$

In life cosmic-ray-produced C^{14} atoms are assimilated by organisms at just the rate that C^{14} atoms disintegrate to form N^{14}. At death of the organism, the assimilation process stops abruptly. Carbon 14 can no longer enter the body of the organism. Radioactive decay takes over in an uncompensated manner. According to the principle of radioactive decay, half of the specific C^{14} radioactivity that was in the organism when it was alive will be present after 5,600 years. After another 5,600 years half of the remaining specific C^{14} radioactivity will be present, and so on. Measurement of this exponential rate of decay provides the built-in chronometer for measurement of time.

Thus, if organic material is present in sediments or is associated with land surfaces, benchmarks in time can be established. By employing the principles of relative dating to the radiocarbon sites, chronology may be developed for the entire system of landscape, sediments, and soils. We shall proceed to develop such systems mainly in the state of Iowa, but it may be necessary to wander farther afield, as we have done in Hawaii, in order to illustrate methods and techniques. Since our emphasis will be on dating by radiocarbon, one should not overlook the fact that only an infinitesimal part of geologic time is involved. By other isotopic-dating methods, the beginning of the life record of the earth has been measured at some 600 million years (Kulp, 1961). Our chronology will be 1/20,000 of that time.

CHAPTER 2

BENEATH THE SURFACE

A LANDSCAPE OR LANDFORM IS THE exterior or external form of a part of the earth's surface, and this is what the eye sees when a land surface is viewed. Something holds up the landscape. Something is beneath the surface, and a considerable part of the physiographic history of an area or a region is contained beneath the surface. The subsurface involves the study of stratigraphy which is the field of geology that deals with the origin, composition, distribution, and succession of strata. Subsurface study is made easy where natural or artificial cuts have been made in the earth. Along a natural cut such as a stream bank, numerous strata may be observed. Artificial cuts are made by man in building roads, railroads, or even in foundation excavations for buildings. If cuts are not available, then the subsurface must be probed by drilling and preferably by the extraction of undisturbed, continuous drill cores. This work is tedious, time-consuming, and hard. Observation is restricted to the sites of drilling, and subsurface relations are extrapolated from one site to another. Cuts, however, permit observation and measurement of a vertical, continuous panorama, but only along a plane or along one direction geographically. Drilling or coring away from the cut allows construction of the third dimension of the subsurface.

Subsurface studies result in determining the succession of strata. For any given region a *generalized stratigraphic section* may be composed which shows the vertical arrangement of strata if all are present. This is the arrangement for the region as a whole. It does not mean that at each place in the region all of the units are present. Some units from the lower, middle, or upper parts of the generalized section may be missing in local areas.

As our major area of interest is the state of Iowa, a generalized section of the Quaternary will include the strata of the Pleistocene and the Recent. The Pleistocene usually connotes glaciation because repeated glaciation was the outstanding event of that time (Flint, 1957). The Recent can be used as the time since the retreat of the last ice sheet (Kay, 1931).

GENERALIZED STRATIGRAPHIC SECTION OF THE QUATERNARY OF IOWA

The present region of Iowa was glaciated four major times during the Pleistocene. These major episodes from oldest to youngest are named Nebraskan, Kansan, Illinoian, and Wisconsin. Between these glaciations were interglacial times when the land surface was not covered by glacier ice and the landscape was worked on by processses of erosion, deposition, weathering, and soil formation such as are effective today. The interglacial units from oldest to youngest are Aftonian, Yarmouth, and Sangamon. The Recent, following Kay's definition, began after Wisconsin glaciation and continues to the present. The generalized stratigraphic section of Iowa may be arranged (Table 2.1). This section is the one that is accepted as standard for all of the midcontinent region of North America.

The importance of work in Iowa in establishing the standard section is shown by the two lower glacial and interglacial stages. Although the Nebraskan is named for the state

TABLE 2.1. GENERALIZED STRATIGRAPHIC SECTION OF THE QUATERNARY OF IOWA

Unit				
Glacial	Interglacial	Type locality or region	Named by	Reference
	Recent	...	...	...
Wisconsin		State of Wisconsin	T. C. Chamberlin	*Jour. Geology*, v. 3, pp. 270–77, 1895
	Sangamon	Sangamon County, Illinois	F. Leverett	*Jour. Geology*, v. 6, p. 176, 1898
Illinoian		State of Illinois	F. Leverett	*Jour. Geology*, v. 4, p. 874, 1896
	Yarmouth	Yarmouth, Iowa	F. Leverett	*Jour. Geology*, v. 6, p. 239, 1898
Kansan		Northeast Kansas	T. C. Chamberlin	*Jour. Geology*, v. 3, p. 271, 1895
	Aftonian	Afton Junction, Iowa	T. C. Chamberlin	*Jour. Geology*, v. 3, p. 270, 1895
Nebraskan		Eastern Nebraska	B. Shimek	*Geol. Soc. Amer. Bull.*, v. 20, p. 408, 1909

of Nebraska, the work was really done on this glacial deposit in Harrison, Monona, and Pottawattamie counties, Iowa. The reasoning in naming the drift "Nebraskan" was somewhat as the following: The drift here on the east side of the Missouri River Valley certainly crosses into Nebraska. Therefore, it is named Nebraskan. The type area in fact is in the three western Iowa counties (Shimek, 1910).

Similar reasoning was used in naming the Kansan drift. The application of the name was actually done in the Afton Junction-Thayer area in Union County, Iowa. The term "Kansan" was applied because of the great extent of the glacial drift in the direction of the arid plains and because the drift in the state of Kansas was free from complications with other glacial deposits. The Afton Junction-Thayer area is the true type locality of the Kansan drift.

This area is also the type locality of the Aftonian, the lowest interglacial stage of the standard section. In fact, Nebraskan, Aftonian, and Kansan were finally conclusively separated in this area.

The Yarmouth interglacial stage was established in a hand-dug well in the village of Yarmouth in Des Moines County in southeastern Iowa. Being in a well, the original section has long been inaccessible, but other sections in the area serve the same purpose.

Thus, four of the major units of the standard glacial and interglacial stages were established in Iowa and are well known around the world. Correlations from other areas, regions, countries, and continents are made to these deposits and events in Iowa.

Again in emphasis, the standard section is very generalized. At no one place in Iowa has the entire column been identified with one stratum upon another vertically. Then too, at any one place the deposits and recognizable events are much more complicated than the generalized section indicates. These complications will be introduced as needed.

In the dating of these strata, all of the deposits beneath the Wisconsin are beyond the datable reach of the radiocarbon method. Currently, the oldest dated sample in Iowa is a piece of spruce wood from a terrace alluvium along the Boyer River at Logan in Harrison County. The age is 37,600 ± 1,500 years (W-880). (All radiocarbon dates will be identified by a sample number such as this one, W-880. A narrative description for each sample site, its location, and other comments about it are in Chapter 6, Catalog of Radiocarbon Dates in Iowa. Reference back to the catalog may be made, if so desired. All of the locations are on plate 1.) More than a dozen other dates in Iowa are "greater than," > 29,000 to > 40,000 years. In the jargon, these samples are "dead." All of them come from deposits older than the Wisconsin.

How much older is "older than the Wisconsin"? In the present state of the art, the absolute ages of Sangamon to Nebraskan are conjectural. One estimate places the beginning of the Nebraskan, that is, the beginning of the Pleistocene, at about one million years ago (Kulp, 1961). This may or may not be a reasonable, educated guess.

Recently, measurements of older age have been made through combined study of radioisotopic, chemical, and microfossil analysis of deep-sea cores of globigerina-ooze. Globigerina are kinds of foraminifers, that is, very small marine organisms having calcareous shells which ultimately form the bulk of sediments on the ocean bottoms. Age evaluation of earlier Pleistocene time follows a round-about method.

The temperature dependence of the oxygen-isotope fractionation in the chemical system of carbon dioxide-water-calcium carbonate is a basis for determining the temperature of precipitation of the carbonate by measuring its oxygen-isotope composition (Emiliani, 1966). The O^{18}/O^{16} ratio in any calcium carbonate sample can be measured with an analytical error of ± 0.1 per mil. A corresponding error in temperature would be ± 0.5° C. Oxygen-isotope composition of surface seawater varies from 0 per mil at the equator to +1 per mil in the tropical evaporation belts to −1 in the high latitudes. The total range of 2 per mil is equivalent to a temperature change of 9° C. Thus, the carbonate shells of certain kinds of foraminifers that live in the near surface waters of the seas have a built-in thermometer. These shells accumulate in sediments on the sea bottom so that a vertical temperature record is preserved in the vertically aligned sediments.

Ages of the shells are determined either by the radiocarbon method or by protactinium-231/thorium-230 (Pa^{231}/Th^{230}) measurements. Temperature variations through time when plotted are like a sine curve with an amplitude of 9° to 10° C and an average wave length of about 50,000 years. In the past 425,000 years eight temperature cycles have approximately the same amplitude. Cold cycles are related to glaciation, and warm cycles are related to interglacials and the Recent. By manipulating all of these data, the temperature-time curve that represents all of the Pleistocene as classically understood goes back to about 425,000 years (Emiliani, 1966). Extrapolating this value to the midcontinent and Iowa would suggest that the beginning of the Nebraskan was about that time. However, this is a considerably wide extrapolation and must be considered as such. The ocean basins not only are far afield but far at sea from the type areas of Iowa.

Here we can determine that certain strata are older than an absolute age that marks the base of a younger stratum. Below this level only the relative ages of things can be determined. However, in general, the older strata can be as old as a half to one million years, but this dating is not very good.

WISCONSIN LOESS IN IOWA

Two major deposits of Wisconsin age are at and beneath the land surface in Iowa (pl. 1). One of them, the Wisconsin loess, covers about 66 percent of the state, or 37,180 of the 56,280 square miles. The other of them, the Wisconsin drift of Cary age, extends from the Minnesota state line southward in a broad lobe to the city of Des Moines. This last feature long has been known as the Des Moines drift lobe. It covers about 22 percent of Iowa, or 12,300 of the 56,280 square miles.

Dating of deposits and landscapes can be keyed to the loess and for many reasons. (1) It has the greatest areal distribution, buries many other deposits, and serves as a lid on them. Consequently, the other deposits are older. (2) The loess, in turn, passes under the drift of the Des Moines lobe which must be younger. (3) The loess has unique internal characteristics that permit its identification readily whether it is at the surface or buried. (4) The loess has been radiocarbon dated at many places at its base, within it, and at its top, so that the loess deposition episode is accurately known to extend from 29,000 to 14,000 years ago.

All of these reasons show that an observer can work downward or upward from the loess and relate other deposits or features to it. This procedure will be followed by starting in southwestern Iowa and working counterclockwise across and around the state to northwestern Iowa. Some digression may be necessary to introduce other relations in context. But before relating other things to the loess, some explanation is needed about the loess itself.

What Is Loess?

Loess is a wind-deposited sediment that is commonly unstratified and unconsolidated and is composed dominantly of silt-size particles. Particle size is measured by the settling of a grain a specific distance in a column of fluid of specific viscosity in a specified time. This type of measurement is used for determining silt and clay sizes. Sand sizes are usually determined by sieving particles through a stack of sieves whose wire-mesh openings decrease in size downward. On any sieve those particles will be trapped whose diameter is slightly greater than the mesh size of this sieve and whose diameter is

TABLE 2.2. PARTICLE-SIZE FRACTION CLASSIFICATION

	Equivalent diameter	
Class	mm	microns (μ)
Gravel	>2.0	>2000
Very coarse sand	2–1	2000–1000
Coarse sand	1–0.5	1000–500
Medium sand	0.5–0.25	500–250
Fine sand	0.25–0.125	250–125
Very fine sand	0.125–0.062	125–62
Very coarse silt	0.062–0.031	62–31
Coarse silt	0.031–0.016	31–16
Medium silt	0.016–0.008	16–8
Fine silt	0.008–0.004	8–4
Very fine silt	0.004–0.002	4–2
Clay	<0.002	<2

slightly smaller than the mesh size of the above sieve. The size fractions are separated by sieving or settling in fluid, and each size is reported as a percent by weight per total sample weight.

The equivalent diameters of size fractions that designate sand, silt, and clay vary with the field of science. The sand-to-silt break is 0.062 mm for geologists but 0.05 mm for pedologists. The silt-to-clay break is 0.005 mm for engineers, 0.004 mm for geologists, and 0.002 mm for pedologists. These differences must be considered in using data published in the various fields. A system of compromise between geology and soils that is used in Iowa is given in Table 2.2. Maximum clay size is 0.002 mm or 2 microns (μ). The remainder is essentially the Wentworth size classification with the addition of the very fine silt grade. This explanation of particle size is

FIG. 2.1. Wisconsin loess in Iowa. (A) On the preexisting rolling Sangamon surface on Loveland (Illinoian) loess in Pottawattamie County. (B) Vertical section below ridge crest in near foreground of (A) showing: Marshall soil at top (1), Wisconsin loess (2) with basal soil (3), Sangamon paleosol (4), and Loveland loess (5). (C) Loess (2) buried beneath Cary glacial till (1) and in turn burying Kansan till (3) near Scranton, Greene County. (D) Loess (2) buried beneath Cary till (1) and in turn burying Kansan till and its paleosol (3) in road cut at Keosauqua Boulevard interchange at freeway in Des Moines.

B
1
2
3
4
5
A
C
1
2
3
D
1
2
3

important in understanding some of the internal properties of the loess.

The Wisconsin loess was not deposited on a level or flat preexisting land surface. Instead, the older surface had valleys, terraces along the valleys, and ridges between the streams (Fig. 2.1). In fact, the interiors of 54 consecutive ridges were exposed during construction of the Rock Island Railroad between Bentley, Pottawattamie County, and Atlantic, Cass County, in 1953 and 1954. In order from west to east, this route crosses Mosquito Creek and the West and East Nishnabotna rivers and their tributaries (pl. 1). A staircase of ridges with intervening valleys rises from each main valley to a divide ridge and then descends to the next main valley. For example, between Indian Creek and East Nishnabotna River which is 5½ miles to the east, eight ridges are crossed. The base of the Wisconsin loess in order is 1217, 1277, 1299, 1294, 1275, 1260, and 1246 feet above sea level. In a terrace along the East Nishnabotna River the base of the loess is at 1110 feet. Thus, the loess was deposited on a dissected surface whose relief was 189 feet. Along the entire route the minimum base elevation of loess is 1100 feet and the maximum is 1318 feet. The lowest value is in a terrace along the West Nishnabotna River at Hancock, Iowa, and the highest elevation is in the divide ridge between the West and East Nishnabotna rivers. The maximum difference in elevation of the surface on which the Wisconsin loess was deposited is 218 feet.

Eastward in northwestern Adair County between Turkey Creek and Middle River, ten ridges are crossed and the base of the Wisconsin loess is as low as 1336 feet and as high as 1420 feet. The high point is the ridge at the Adair City Park, and this ridge is the Missouri-Mississippi River divide at this latitude in Iowa. The local maximum relief at the base of Wisconsin loess in these ten ridges is 84 feet, but the maximum difference from Bentley through Adair is 320 feet.

Herein is prime evidence for the eolian (wind) origin of the loess. The silts blanket all ridges regardless of their elevation. There is only one common thing above all ridges, the atmosphere, and dust had to settle out of it to form a capping on all ridges. Two other properties of the loess, thickness distribution and particle-size distribution, fit within this eolian framework and are taken up in order.

By numerous measurements of thickness of the Wisconsin loess in road cuts and in drilling, a contour map may be constructed (Fig. 2.2). Three major provinces are noted. (1) Loess is more than 64 feet thick adjacent to and east of the

FIG. 2.2. Loess thickness in Iowa. Contours in feet. Note decrease in thickness away from the Missouri River Valley and around the periphery of the Iowan erosion surface in northeast Iowa (cf. pl. 1). Shaded areas are essentially loess-free. (Modified from Thorp and Smith, 1952.)

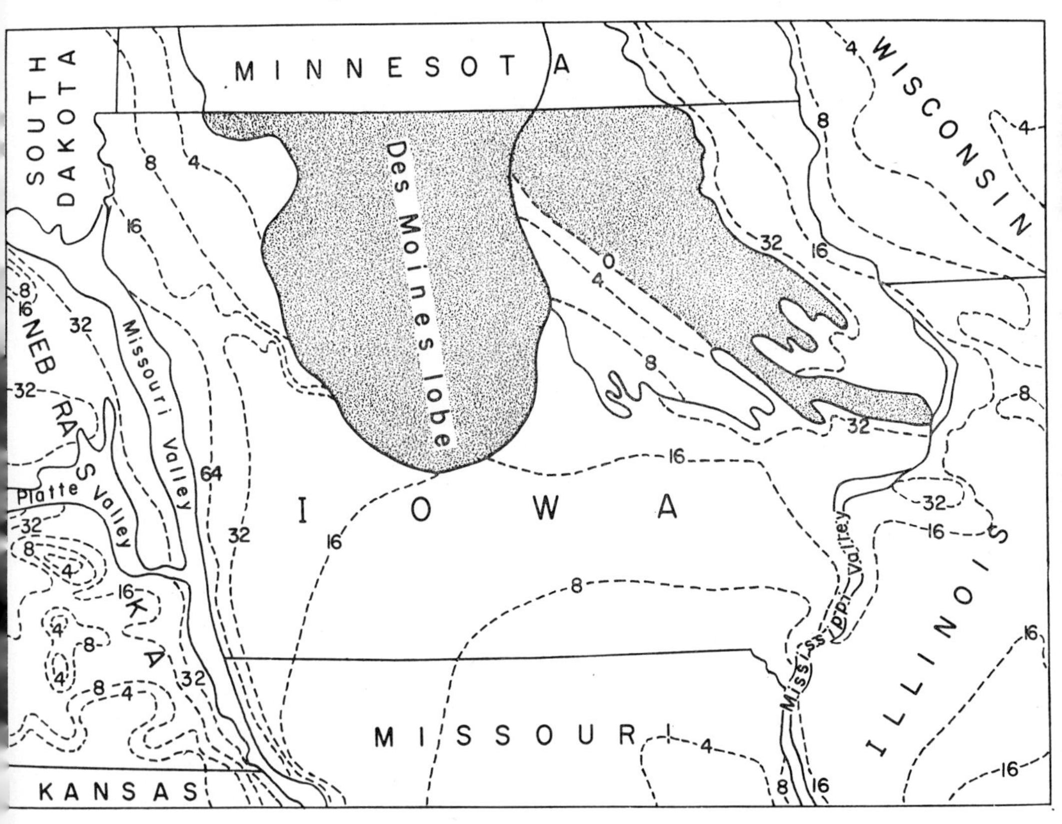

Missouri River Valley and thins away from the valley. (2) Loess is more than 32 feet thick around the margin of the Iowan erosion surface in northeast Iowa and thins away from the margin. (3) The Des Moines drift lobe and a major part of the Iowan erosion surface in northeast Iowa are essentially loess-free.

Direction of loess thinning is indicated by any continuous line that crosses the thickness contours at right angles. For example, in northwest Iowa the direction of thinning is to the northeast. Starting at the Missouri Valley in western Iowa, the direction of thinning is to the southeast.

The thinning of the loess with distance along a direction is not haphazard but extremely systematic. It is so systematic that the relation can be expressed mathematically. If a traverse is started in Monona County, east of the Missouri River Valley, and extended to the southeast to Wayne County (Fig. 2.2, cf. pl. 1), the loess thickness decreases from nearly 700 inches to slightly less than 100 inches in a distance of about 170 miles. This relation is expressed by the equation

$$Y = 1250.5 - 528.5 \log X$$

where Y is the loess thickness in inches and X is the distance in miles (Hutton, 1947). This mathematical equation simplifies and organizes a lot of measurements so that the relations between loess thinning and distance along a traverse are readily understood. (See Part II for a brief explanation of the methods of calculating the mathematical relations involved here as well as in other cases to be presented later.) This equation means that the rate of thinning of the loess progressively decreases per equal unit of distance along the traverse. From another view, it means that for equal units of decrease in loess thickness, progressively greater segments of distance are required. In other words, the relation may be pictured as a curve that descends and progressively flattens as the distance increases.

Along the Rock Island Railroad route the loess thins to the east as expressed by the equation

$$Y = 1/(9.51 \times 10^{-4} + 7.99 \times 10^{-5} X)$$

where Y is loess thickness in inches and X is distance in miles (Fig. 2.3). This equation appears more complicated but it is

FIG. 2.3. Relations of loess thickness, particle size, and carbonate content of Wisconsin loess at primary- and secondary-divide ridge crests along the Rock Island Railroad from Bentley to Adair, Iowa.

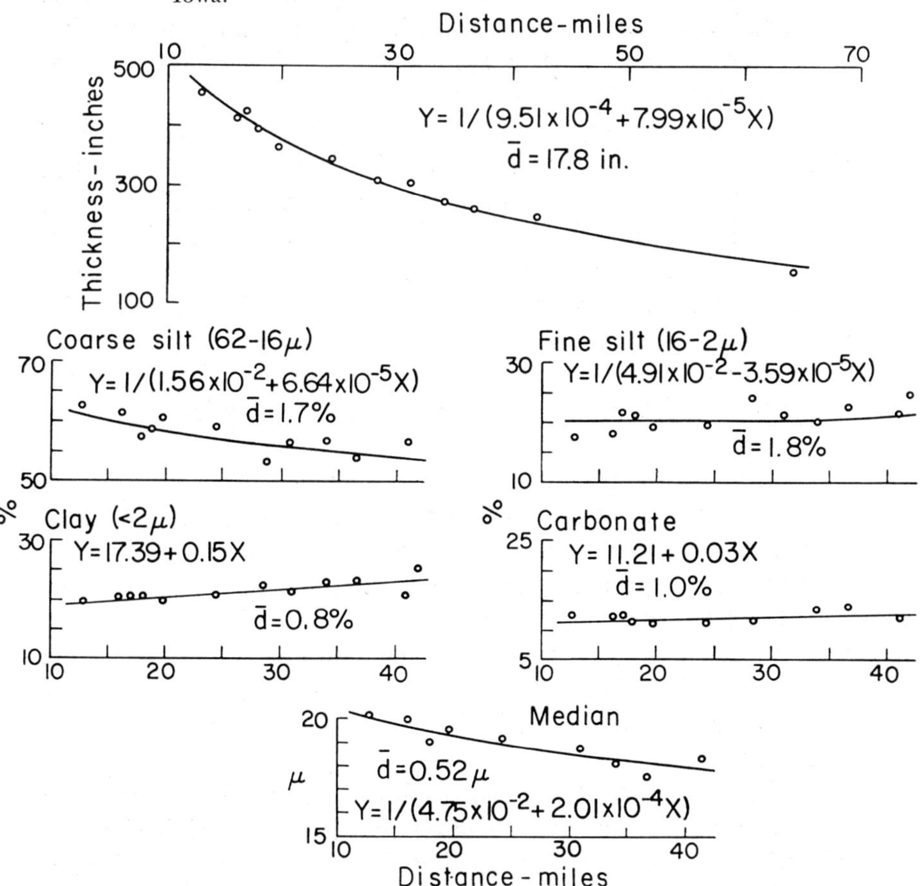

not. In general form it is $Y = 1/(a + bX)$ which will be recalled from the analytic geometry as representing the curve of a hyperbola. In other words, this equation also means that the loess progressively thins from west to east and that the rate of thinning progressively decreases per equal unit of distance along the way.

This last example was introduced for three reasons: (1) It illustrates a common kind of decrease of loess thickness with increase in distance, but that actual changes may be expressed differently in mathematics. (2) It permits a direct comparison of this thinning system with the topography along the route that has been previously explained. (3) It allows a comparison to be made of the thinning system to other internal properties of the loess that were measured along the route.

In this thinning system the thickest loess is just east of the Missouri River Valley, and thinning takes place to the northeast, the east, and the southeast (Fig. 2.2). This shows that the big valley was the source of the silts. Prevailing westerly and northwesterly winds picked up the silts from channel bars and floodplains of the river and carried them in the atmosphere in the easterly direction. Greater amounts fell from the atmosphere to the east and near the valley than progressively farther away, even to Wayne County in south central Iowa.

The thinning was not affected by the topography over which the silts were blown and on which they were deposited. The thinning system that can be described mathematically crosses the ridges along the Rock Island route that stand at elevations through a range of 320 feet. As pointed out, these ridges go up and down like a series of staircases, yet the loess thins across them from west to east. It even thins uninterruptedly across the Missouri-Mississippi River divide at the town of Adair.

This directional change in thickness is substantiated by and coupled with a systematic change in the particle-size distribution in the loess. Recall that this discussion opened with particle size, and the discussion will close with it. The coarser particles of the silt decrease in amount as the loess thins from west to east across the rough topography along the Rock Island route (Fig. 2.3). The amount of finer silt particles increases. The amount of clay-size particles increases. The median diameter which is the midpoint size of all particles decreases. The decrease of coarser silt and median diameter with distance is expressed by the curve of a hyperbola which is alike in kind to the loess thickness curve. The finer silt has the same kind of curve but with a reversed slope to it. The increase in clay sizes is a direct straight-line relation (Fig. 2.3). All of these features require a selective sorting of sizes of particles by some medium such as water or wind. The large size regional pattern points to wind alone as the responsible agent.

These internal properties of the loess are very important in the nature of soils that formed on the loess surface. Reference will be made to these properties in later discussion.

Dating the Loess

Since the loess was deposited on a previous land surface, the first thin increments of the silts became incorporated in and on the organic horizons of the soils on that land surface. In depressions on that surface, peats had even formed. Regardless of whether the prior organic horizons were A or O horizons (see Background), the organic matter, including wood fragments in those horizons, may be radiocarbon dated. Where dates are determinable, the age of the base of the loess is outlined. Such dating has been done at many places (Table 2.3).

The base of the Wisconsin loess in Iowa ranges in age from 16,500 to 29,000 years, but this broad spectrum of dates does not show any landscape system in the simple chronological listing. If the loess traverse from the Missouri River Valley to south central Iowa is laid out and dates are fitted, age relations can be better understood. In sequence the basal dates are 23,900 (I-1420), 24,500 (W-141), 18,700 (I-1411),

TABLE 2.3. RADIOCARBON DATES AT BASE OF WISCONSIN LOESS

Sample*	Date in years before present	Location†	Notes
I-1419A	16,500 ± 500	Humeston, Wayne County	Soil organic carbon (OC), residue
I-1020	17,030 ± 500	Des Moines, Polk County	OC, residue‡
W-1687	18,300 ± 500	Salt Creek, Tama County	OC, residue
I-1411	18,700 ± 700	Greenfield, Adair County	OC, residue
I-1419B	19,000 + 6,000 − 3,000	Humeston, Wayne County	Humic-acid fraction of I-1419A
W-879	19,050 ± 300	Logan, Harrison County	Spruce wood
I-1408	19,200 ± 900	Harvard, Wayne County	OC, residue
I-1022	20,290 ± 1,000	Kinross, Keokuk County	OC, residue‡
I-1409	20,300 ± 400	Hayward paha, Tama County	OC, residue
I-2332	20,700 ± 500	Alburnett paha, Linn County	OC, residue
I-1410	20,900 ± 1,000	Murray, Clarke County	OC, residue
I-1023	23,900 ± 1,100	Bentley, Pottawattamie County	Spruce wood
W-1681	21,600 ± 600	Palermo area, Grundy County	OC, residue
I-1404	22,600 ± 600	Palermo area, Grundy County	OC, residue
I-1420	23,900 ± 1,100	Bentley, Pottawattamie County	OC, residue
I-1403	23,900 ± 1,100	Grinnell, Poweshiek County	Peat, conifer zone
W-141	24,500 ± 800	Hancock, Pottawattamie County	Larch wood from peat
I-1406	24,600 ± 1,100	Kinross, Keokuk County	OC, residue
I-1267	25,000 ± 2,500	Hayward paha, Tama County	OC, residue
I-1269	29,000 ± 3,500	Salt Creek, Tama County	OC, residue

* Sample numbers are I for Isotopes, Inc.; and W for U.S. Geological Survey, Washington, D.C.

† See Radiocarbon Catalog, Chapter 6, for details.

‡ Organic carbon converted to strontium carbonate.

20,900 (I-1410), 16,500 (I-1419A), 19,200 (I-1408). These values are taken from Table 2.3 and the locations of the samples are on plate 1. These sample sites are at distances of 42, 53, 109, 128, 155, and 158 miles along the traverse. The age values zig and zag up and down with distance, but a trend analysis of the values is

$$Y = 26{,}500 - 55X$$

where Y is the age of the base of the loess in years and X is the distance in miles. The predicted mean age of the base of the loess near the Missouri River Valley is about 26,500 years and in south central Iowa is about 18,000 years. The base of the loess becomes progressively younger toward the east within the loess distribution pattern that is dependent upon the Missouri River Valley as a source. The A and O horizons of the buried soil marking the base of the loess transgress 8,500 years of time.

This time transgression introduces another tool, the *soil-stratigraphic unit,* that is useful in relating the ages of deposits and surfaces. A soil-stratigraphic unit is a soil with physical features and stratigraphic relations that permit its consistent recognition and mapping as a stratigraphic unit (Richmond and Frye, 1957; Amer. Comm. Strat. Nom., 1961). The soil is formed from underlying materials which may be of diverse composition and age. It is defined on the basis of observable features and stratigraphic relations at a locality and may be extended as far as it can be recognized. It may parallel or transgress time.

In applying these requirements to the basal soil of the Wisconsin loess in Iowa, difficulties arise in previously applied classification. This specific soil was believed to separate a lower increment of the loess called Farmdale from the remainder of the overlying Wisconsin loess (Ruhe, 1954). The Farmdale in the area where it was defined in Illinois is 22,000 to 28,000 years old (Frye and Willman, 1960). In Iowa the age range as demonstrated is 16,500 to 29,000 years, which not only includes the defined Farmdale of Illinois but also that part of the Wisconsin formerly called Tazewell which is 17,000 to 22,000 years old. Consequently, the term Farmdale is no longer applied to the basal soil of the Wisconsin loess in Iowa. The informal term of basal soil is adequate.

Locally, wide divergence in age may exist in the basal soil-landscape. In Tama County, Iowa, the age of the A horizon of the soil on a buried hilltop is 29,000 years (I-1269). Only 650 feet away and on a buried surface that is 11 feet lower, the age of the soil is 18,300 years (W-1687). At another near locality, the buried hilltop soil is 25,000 years old (I-1267). Only 400 feet away and 7 feet lower, the age of the buried hillslope soil is 20,300 years (I-1409). In Linn County (Ruhe *et al.,* 1968) the hilltop soil buried under Wisconsin loess is 20,700 years old (I-2332). On an adjacent hillslope 1,200 feet away and 35 feet lower, the soil that is buried be-

TABLE 2.4. RADIOCARBON DATES FROM WITHIN AND AT TOP OF THE WISCONSIN LOESS

Sample*	Date in years before present	Location	Notes
Top of loess and base of Cary till			
W-513	13,820 ± 400	Scranton No. 1, Greene County	Spruce wood
W-517	13,910 ± 400	Scranton No. 2, Greene County	Spruce wood
I-1402	14,200 ± 500	Nevada, Story County	Spruce wood
W-512	14,470 ± 400	Scranton No. 1, Greene County	Fir, hemlock, larch, and spruce wood
W-153	14,700 ± 400	Clear Creek, Story County	Hemlock wood
Within loess			
I-1270	16,100 ± 1,000	Boone, Boone County	Spruce wood
I-1024	16,100 ± 500	Madrid, Polk County†	Spruce wood
C-528	16,367 ± 1,000	Clear Creek, Story County	Hemlock wood
W-126	16,720 ± 500	Mitchellville, Polk County	Yew, spruce, and hemlock wood
C-481	>17,000	Mitchellville, Polk County	Yew, spruce, and hemlock wood

* Sample numbers are I for Isotopes, Inc.; W for U.S. Geological Survey, Washington, D.C.; and C for University of Chicago. All Chicago dates are by original carbon black method.

† Madrid is in Boone County, but location is just across county line in Polk County.

neath alluvium derived from the loess is 12,700 ± 290 years old (I-2333). In this last case the soil-stratigraphic horizon transgresses even younger Wisconsin time, that of the Cary which is 13,000 to 14,000 years ago in Iowa. In fact, it even includes a part of postglacial time in Iowa, as will be seen later.

Other dates at the base of the Wisconsin loess are in other places in Iowa and may be examined in Table 2.3 and placed in context by referring to plate 1 and the Radiocarbon Catalog in Chapter 6.

Radiocarbon dates have also been determined from within and at the top of the Wisconsin loess (Table 2.4). All

FIG. 2.4. (A) First road cut west of the Des Moines River Valley on new U.S. Highway 30 in Boone County, showing Cary till (1) over Wisconsin loess (2) over pre-Wisconsin paleosol (3) on Kansan till (4). Des Moines Valley is at right margin, looking north. Spruce wood from loess is 16,100 years old. (B) Spruce tree rooted in place at top of Wisconsin loess and beneath Cary till at Scranton No. 1 section in Greene County. Tree is 13,820 years old. (C) Spruce tree rooted in place at top of Wisconsin loess and beneath Cary till at Scranton No. 2 section in Greene County. Tree is 13,910 years old. See this section in Table 2.6.

A
1
2
3
4
B
C

of the sample locations are from road cuts that expose the Wisconsin loess buried beneath the Cary till of the Des Moines glacial drift lobe (pl. 1). The best time to examine these cuts is soon after they have been constructed (Figs. 2.3 and 2.4). Commonly, spectacular features become visible. In the Scranton No. 1 road cut along new U.S. Highway 30 in Greene County, a buried forest was exposed by road construction machinery. As horizontal benches were cut downward, as many as 15 to 20 buried trees were truncated in 50 square feet of area. The tree stumps with their root systems were in place (Fig. 2.4). The kinds of trees were hemlock, larch, fir, and spruce. Wood from one of the trees was 13,820 years old (W-513). A similar kind of feature was exposed in a cut along a county road at Scranton No. 2 in Greene County. There a spruce tree rooted in place was 13,910 years old.

In both of these places the radiocarbon dates give answers to two questions. What is the age of the top of the loess, and what is the age of the base of the Cary till? The trees were growing on the loess surface, hence they give a minimum age for the loess. The trees were knocked over toward the south by the glacier ice that deposited the Cary till, hence they give a maximum age for the Cary till. A simple average of all sections in Iowa that marks the till-loess contact is 14,220 years or rounded to 14,000 years (Table 2.4).

At other localities where the loess is buried beneath the Cary till, logs of coniferous wood have been dated from 16,100 to more than 17,000 years (Table 2.4). When these dates are aligned with those from the base of the loess (Table 2.3), the age of the Wisconsin loess in Iowa is 14,000 to 29,000 years. This age range is based on the reasonable number of 30 samples from many parts of the state.

Fauna of the Loess

Another conspicuous feature of the loess is the great abundance of fossil mollusks in it. The shells of gastropods

FIG. 2.5. Typical faunal assemblages in zones in the Wisconsin loess in Iowa. All species are terrestrial gastropods (land snails) except C which is an aquatic pulmonate gastropod. (From Leonard, 1952.) Vertical white scale to right of A is 2 mm, and applies to all species.

LOWER ZONE

A *Succinea avara*
B *Helicodiscus parallelus*
C *Lymnae parva*
D *Pupilla muscorum*
E *Pupilla blandi*
F *Hawaiia minuscula*
G *Vallonia gracilicosta*
H *Deroceras laeve*

TRANSITION ZONE

I *Succinea grosvenori*
J *Succinea avara*
K *Discus cronkhitei*
L *Discus shimeki*
M *Helicodiscus singleyanus*
N *Hawaiia minuscula*
O *Pupilla muscorum*
P *Deroceras laeve*
Q *Helicodiscus parallelus*
R *Valonia gracilicosta*
S *Pupilla blandi*
T *Vertigo modesta*

UPPER ZONE

U *Succinea ovalis*
V *Discus shimeki*
W *Hendersonia occulta*
X *Vallonia gracilicosta*
Y *Pupilla muscorum*
Z *Deroceras laeve*
AA *Columella alticola*
BB *Vetrigo modesta*
CC *Helicodiscus singleyanus*
DD *Cionella lubrica*
EE *Succinea grosvenori*
FF *Discus cronkhitei*
GG *Helicodiscus parallelus*
HH *Pupilla blandi*
II *Retinella electrina*
JJ *Hawaiia minuscula*
KK *Striatura milium*

or snails are so plentiful that they comprise a large part of sediments at some places. The fossils greatly exceed in variety of species and in abundance the local living snail population. The greater part of the fossil fauna is either extinct or no longer living in the region. These last two facts go to the heart of the matter. The fossils can be used for stratigraphic purposes in tracing the loess where it is the deposit beneath the land surface to where it is a deposit buried under younger deposits. Most of the snails are terrestrial gastropods, that is, they lived on land. A few are pulmonate gastropods of aquatic habit. Both kinds are air breathers (Fig. 2.5). The kinds of fossils are another line of evidence to add to those previously introduced to show the wind-blown origin of the loess. Land snails lived on the land surface and had to be buried by sediment deposited from above them. On the hill summits, there was only one possible source and that was from the atmosphere. If water of any kind had been involved, the fauna should be dominantly aquatic, but it is not.

Does the stratigraphic use of the fossil snails work? Yes, and this can be demonstrated by returning to the sites of the loess traverse in southwestern Iowa (Fig. 2.3). West of the West Nishnabotna River two faunal zones are common, but east of the river only one zone is present (Table 2.5). Now if thicknesses of loess are measured to bases of the faunal zones, the ages of parts of the loess can be determined relative to the zones. For example, in cut 50 the total thickness of loess is 453 inches. To the bases of the upper and lower faunal zones in the loess, the thicknesses are 190 and 274 inches, respectively. The upper 30 percent of the loess is younger than the upper faunal zone, and the upper 60 percent is younger than the lower faunal zone. Similarly, along the entire traverse 60 to 65 percent of the loess is younger than the lower faunal zone. Absolute dating of the fauna would put the whole system in business.

Does the stratigraphic use of the fossil snails work where the loess is buried beneath the Cary till of the Des Moines drift lobe? Yes, and this can be demonstrated by comparing a sec-

TABLE 2.5. FAUNAL ASSEMBLAGES IN WISCONSIN LOESS ALONG TRAVERSE IN SOUTHWESTERN IOWA

Species and zones of Leonard, 1952	Railroad cuts and faunal zones in cuts									
	50		45		39	33		25	17	11
	Upper	Lower	Upper	Lower	Upper	Upper	Lower			
Upper zone										
Succinea ovalis	x		x		x	x				
Hendersonia occulta	x		x			x				
Retinella electrina	x									
Columella alticola			x		x	x				x
Upper and transition zone										
Succinea grosvenori	x	x	x	x	x	x	x	x	x	x
Discus cronkhitei	x	x		x			x	x	x	x
Discus shimeki		x	x	x	x	x	x	x		x
Vertigo modesta				x			x			x
Lower and transition zone										
Vallonia gracilicosta	x	x	x							
Pupilla blandi	x				x			x	x	x
Hawaiia minuscula				x	x			x		x
Euconulus fulva								x		x
Lower zone										
Succinea avara	x	x						x	x	x
Lymnaea parva								x		

x—Indicates presence.

tion, cut 33, where the loess is beneath the land surface, to the sections at Scranton No. 1 and No. 2 in Greene County and Clear Creek in Story County where the loess is buried beneath Cary till (Table 2.6). The faunal assemblages are very similar, yielding another line of evidence that the loess indeed passes beneath the Cary till.

Since the radiocarbon dates at the till-loess contact are about 14,000 years, this minimum age can be assigned to the upper faunal zone of the loess. Consequently, beyond the Cary drift lobe, where the Wisconsin loess is at and beneath the land surface, the top of the upper faunal zone is older than 14,000 years.

Alteration of the Loess

Weathering profiles (see Background) are prominent in the loess throughout Iowa. Beneath the surface are six zones. (1) The *oxidized and leached zone* has colors of yellowish brown in the sediment matrix. Light gray mottles are commonly present. Carbonates are not present, so the sediment does not effervesce when a drop of hydrochloric acid is applied. (2) The *oxidized and unleached zone* does have carbonates and does effervesce upon acid application. Commonly, grotesquely shaped carbonate-cemented nodules are present and are known as "loess kindchen or loess dolls." Colors of the sediment are similar to (1). (3) The *deoxidized and leached zone* has a light gray sediment matrix in which iron oxide tubules are oriented vertically. The tubules are reddish brown and commonly have concentric reddish-brown bands with intervening gray bands around them (Fig. 2.6). These tubules are known as "pipestems." This zone does not have carbonates and does not effervesce with acid. (4) The *deoxidized and unleached zone* has similar features to (3), but in addition has carbonates, does effervesce, and has loess kindchen. (5) The *unoxidized and leached zone* has sediment matrix colors of dark gray, blue, bluish green, and green. It does

TABLE 2.6. FAUNAL ASSEMBLAGES IN WISCONSIN LOESS BEYOND AND BENEATH THE CARY TILL

	Sections and faunal zones in cuts																	
	Cut 33						Scranton No. 2			Scranton No. 1			Clear Creek					
	Lower zone			Upper zone									Lower zone			Upper zone		
Species	L	T	U	L	T	U	L	T	U	L	T	U	L	T	U	L	T	U
Columella alticola						x			x			x						x
Euconulus fulvus									x			x						x
Hendersonia occulta						x			x									x
Succinea ovalis						x			x			x						x
Discus cronkhitei		x	x					x	x		x	x		x	x		x	x
Discus shimeki		x	x		x	x					x	x						
Succinea grosvenori		x	x		x	x		x	x		x	x		x	x		x	x
Vertigo modesta		x	x					x	x		x	x					x	x
Hawaiia minuscula	x	x	x				x	x	x				x	x	x			
Pupilla blandi							x	x	x	x	x	x	x	x	x	x	x	x
Pupilla muscorum							x	x	x				x	x	x			
Vallonia gracilicosta	x	x	x	x	x	x	x	x	x	x	x	x	x	x	x	x	x	x
Succinea avara							x	x		x	x		x	x		x	x	
Lymnaea parva										x			x					

x—Indicates presence; L, T, and U are lower, transition, and upper faunal zones of Leonard, 1952.

FIG. 2.6. Weathering zones of the Wisconsin loess in Iowa. (A) Detail of the deoxidized zone, showing vertically oriented reddish-brown pipestems with concentric banding around them. The sediment matrix is light gray. (B) Deoxidized zone (1) above a reddish-brown paleosol (2). (C) Recent beveling of hillslopes along a drainage net so that the deoxidized zone, light colored, is exposed along hillslope contours in southwestern Iowa.

not have pipestems, does not have carbonates, and does not effervesce. Commonly, organic matter is along bands and layers and in flecks. Wood fragments may be present. (6) The *unoxidized and unleached zone* is similar to (5), but has carbonates and loess kindchen so that the sediments effervesce upon acid test.

All of these zones do not necessarily occur in one vertical sequence or in the order given. Generally, on a flat or slightly rounded ridge the oxidized and leached zone is beneath the surface. Any of the other zones, excluding the unoxidized zones, may be the next zone downward. Where unoxidized zones are present, they are universally just above the base of the loess. The unoxidized and unleached zone is the least altered by weathering processes and is the zone that generally contains the organic samples used in radiocarbon dating.

As mentioned previously, these zones are defined on the chemical basis of the nature and distribution of calcium and magnesium carbonates and iron oxide. The carbonates are detected by the simple acid test and the visual presence of the loess kindchen and fossil shells which are also calcareous. Where the sediments are oxidized, their matrix color is yellowish brown, and iron oxide is diffuse throughout the sediment. Free iron oxide content, defined as the iron chemically extracted with sodium hydrosulfite and expressed as oxide, varies from 1.5 to 3.5 percent. In the light gray matrix of the deoxidized zone, the free iron-oxide content is much less, being 0.3 to 0.6 percent (Daniels *et al.*, 1961). Consequently, during the subsurface weathering, iron has been removed from the sediment matrix at one place, but then has accumulated in pipestems at another place. The free iron-oxide content of the pipestems varies from 9 to 26 percent.

The properties of the deoxidized zone are suggestive of a weathering process called *gleying* which involves chemical reduction in a water-saturated or nearly saturated zone in the presence of organic matter. The organic matter content of these zones still ranges from 0.1 to 0.2 percent. However, ferrous iron presently is only a trace in these zones. Thus, the deoxidized zones probably reflect relict gleying in the subsurface when saturated ground water zones stood higher on the landscape than they do today (Ruhe and Scholtes, 1956).

The manner in which the zones pass through the hills and beneath the present land surface is suggestive of this kind of origin (Fig. 2.7). Contacts between weathering zones parallel the buried land surface on which the Sangamon paleosol formed. The present land surface does not parallel the zone contacts, and the hillslopes even bevel the zones so that they crop out on the surface (Fig. 2.6). Apparently, subsurface zones of water perched on the clayey paleosols and saturated the overlying loess, permitting gleying and the formation of the deoxidized zones. Currently, the zones are many feet above any water-saturated layers. The zones must represent conditions of the past when the environment differed.

When were conditions different during the past? This problem may be attacked by relating the deoxidized weathering zones to the faunal zones along the traverse in southwestern Iowa (Fig. 2.8). The weathering zones are independent of faunal zones (Ruhe and Scholtes, 1955; Ruhe, Daniels, and Cady, 1967). At site 50 the upper deoxidized zone coincides with the *lower* faunal zone, but at sites 45 and 39 the upper deoxidized zone coincides in part with the *upper* faunal zone. At sites 25 and 17 the upper deoxidized zone is again coincident with the *lower* faunal zone. Consequently, the deoxidized zone must represent a condition of weathering alteration that was impressed on the loess after the loess was deposited. The shells of the fossil snails are particles in the silty sediment.

As previously pointed out, loess deposition stopped about 14,000 years ago, so the deoxidized zone formed after 14,000 years. At site 39 the deoxidized zone crops out on hillslopes like the pattern at cut 45 (Fig. 2.7), and the hillslopes descend to a gully fill whose basal organic sediment is radiocarbon dated at 6,800 years (W-235; see Radiocarbon Catalog, Chapter 6). The deoxidized zone had to be formed so that the hillslope could bevel it, and must be older than 6,800 years. Thus, the deoxidized zone formed between 6,800 and 14,000 years ago.

What environment during that time could have produced the higher stands of subsurface water saturation necessary for the formation of deoxidized zones? As will be seen later, from

FIG. 2.7. Subsurface distribution of weathering zones through the cut-45 ridge along the loess traverse in southwestern Iowa. Note the parallelism of the zones to the paleosol surface on Loveland loess, but the lack of parallelism of the present land surface to the zones. Hillslopes actually bevel the zones (cf. Fig. 2.6). Soils are fitted to the landscape and outcrops of zones.

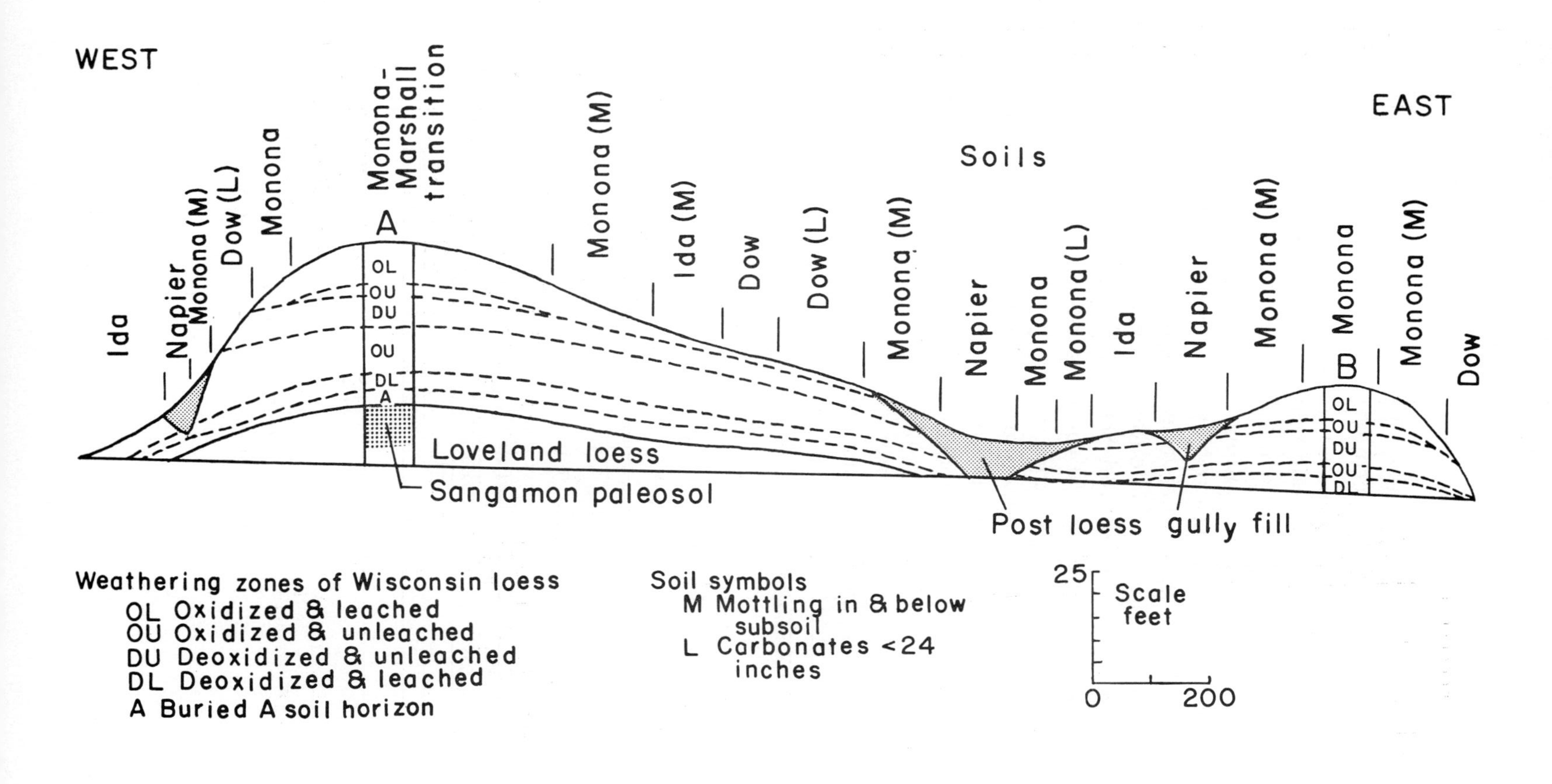
WEST
EAST
Soils
Ida
Napier
Monona (M)
Dow (L)
Monona
A
Monona-Marshall transition
Monona (M)
Ida (M)
Dow
Dow (L)
Monona (M)
Napier
Monona
Monona (L)
Ida
Napier
Monona (M)
B
Monona
Monona (M)
Dow
OL
OU
DU
OU
DL
A
Loveland loess
Sangamon paleosol
Post loess gully fill
Weathering zones of Wisconsin loess
OL Oxidized & leached
OU Oxidized & unleached
DU Deoxidized & unleached
DL Deoxidized & leached
A Buried A soil horizon
Soil symbols
M Mottling in & below subsoil
L Carbonates <24 inches
Scale feet
25
0
200

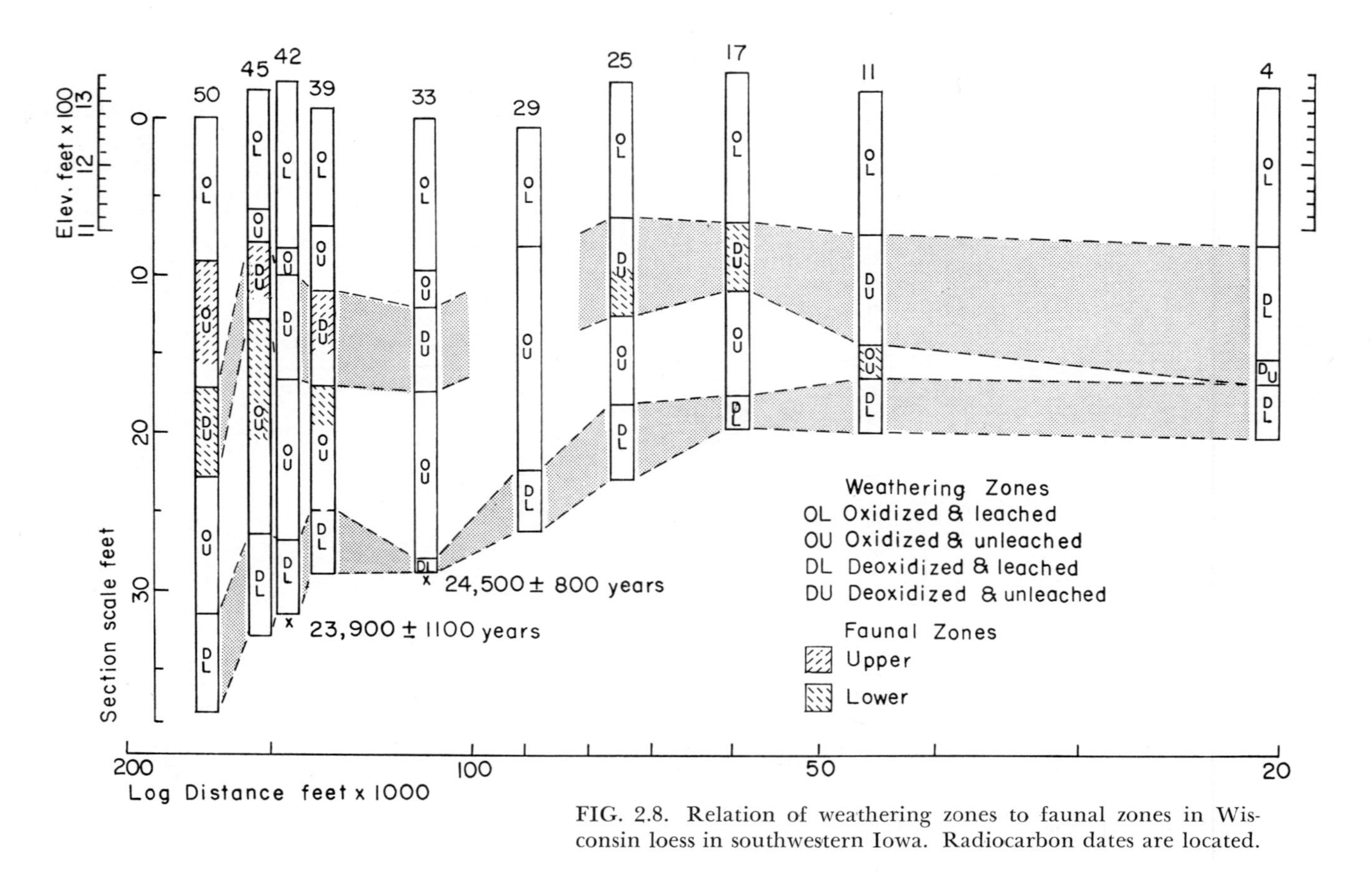

FIG. 2.8. Relation of weathering zones to faunal zones in Wisconsin loess in southwestern Iowa. Radiocarbon dates are located.

13,000 to 14,000 years ago, glacier ice occupied 12,300 square miles of north central Iowa and formed the Des Moines drift lobe. As late as 10,000 years ago and after Cary glaciation, coniferous forest with spruce was present in the state. Both glaciation and coniferous forest indicate a cooler and relatively more moist climate than at present.

The temperature difference may be estimated by comparing present ranges in Iowa where spruce is not a native tree to ranges in northern Minnesota and the upper peninsula of Michigan where spruce is native. Average annual winter temperatures to the north are 0° to 16° F, and annual summer temperatures are 60° to 66° F. In Iowa, values for the same periods are 14° to 24° F and 72° to 76° F. The native spruce areas today are about 11° F cooler than Iowa in both winter and summer. Certainly, during the glacial time of 13,000 to 14,000 years ago and the time following the glaciation, the temperatures in Iowa should have been colder.

The term "relatively more moist" is used for climate for the following reasons. The present annual rainfall in the northern spruce area and in Iowa does not differ greatly, being 22 to 32 inches and 26 to 36 inches, respectively. However, the distribution annually differs radically. About 70 percent of the precipitation in Iowa is during the warm months, but prolonged droughts are common during any season, affecting the general water supply (Reed, 1941). In the northern spruce area, rainfall is fairly well distributed throughout the growing season, and droughts occur occasionally but are not so severe as in Iowa (Wills, 1941). Evaporation is also less than to the south (Hovde, 1941). Consequently, subsurface storage of water is effectively greater in the spruce area. Similar conditions probably existed in Iowa 10,000 to 14,000 years ago.

The deoxidized zones and their related environments have important effects in the formation of soils on the landscape. In this regard, the zones will be reintroduced later.

Since the Wisconsin loess is being used as a key to other deposits in Iowa, another major deposit can be related by moving upward stratigraphically to the Cary glacial drift. Later, it will be necessary to descend again to relate buried landscapes and soils.

CARY GLACIAL DRIFT IN IOWA

The Cary glacial drift is confined to a drift lobe that extends southward 135 miles from the Minnesota state line to the city of Des Moines (pl. 1). The lobe is positioned slightly west of the north-south center line of the state. The Des Moines River flows southward almost along the axis of the lobe. With all of this name association, this large physigraphic feature has long been known as the Des Moines glacial drift lobe. As mentioned previously, the lobe occupies about 12,300 square miles in Iowa. Four major belts of aligned ridges, end moraines, pass around the lobe. Each major belt is set inward and northward from an older system of ridges, and from south to north the end moraines are the Bemis, Altamont, Humboldt, and Algona (Fig. 2.9). These moraines mark the margins of parts of the lobe, and the history of the lobe may be constructed relative to these parts. An understanding of this history, then, involves not only a material but landforms and landscapes associated with the material.

What Are Drift and Moraine?

Glacial drift is all rock material in transport by glacier ice, all deposits made by glacier ice, and all deposits dominantly of glacial origin made in the sea or in bodies of glacial melt water, whether rafted in icebergs or transported in the water itself. Drift includes till and stratified drift (Flint, 1957). *Glacial till* is an unsorted sediment whose particles range in size from boulders to clay. The larger sized particles are randomly dispersed in a matrix composed of sand, silt,

FIG. 2.9. Major end moraines of the Cary drift of the Des Moines lobe in Iowa. The Humboldt moraine bifurcates with one member, the Fort Dodge, extending into northern Webster County, and a second member, the Rutland, crossing Humboldt County.

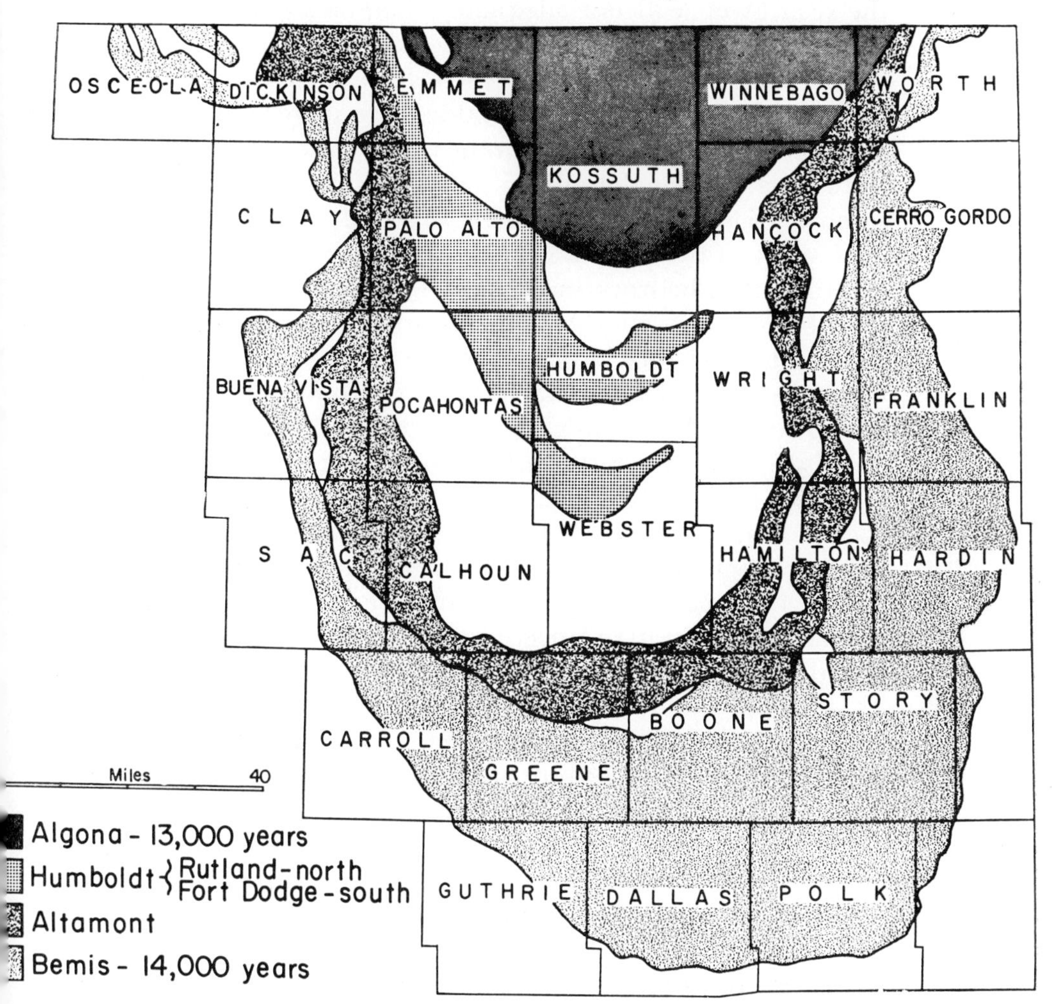

and clay. This random arrangement is nonsorting and shows the lack of water as a transporting agent.

Particle size is again important in identifying the sediment till. Previously (Table 2.2), size classes were introduced in explanation of a fine-textured sediment such as loess, but the sizes given will not adequately handle some of the particles of till. So, larger sizes must be added. Pebble size ranges from 2 to 64 mm (0.1 to 2.5 inches). Cobbles are 64 to 256 mm (2.5 to 10 inches), and boulders are larger than 256 mm (10 inches). In the till of notheastern Iowa, some boulders were so large that they were quarried, converted to building stone, and used for construction purposes.

Many of the particles, including the largest boulders, are not related to bedrock in the area in which they occur in the till. These are *erratics.* Their sizes show that only a transporting agent such as glacier ice would be competent to move them great distances from their source. In southern Iowa Sioux quartzite is common in the till. The nearest outcrop of the rock is the extreme northwestern corner of the state. St. Cloud granite is in the tills all over Iowa. The nearest outcrop of this rock is near St. Cloud in central Minnesota. A few pieces of native copper have been collected from the till in northeastern Iowa. The nearest outcrop of this rare erratic is in the Keeweenawan Peninsula projecting into Lake Superior in Upper Michigan.

Stratified drift is bedded and sorted and shows that water has been active in transporting the sediment. The burden of proof here concerns demonstrating that glacial melt water was the transporting agent. A synonym for stratified drift is *outwash,* but many stratified deposits in Iowa and elsewhere have been termed outwash when they are not. Instead, they are ordinary alluvium that was deposited by water unrelated to glacier ice. Where melt water can be proved, stratified drift confined to valleys forms *valley trains.* Where this drift spreads across the land in a plain, an *outwash plain* results. These two features are the landforms and landscapes formed by melt water deposition in Iowa.

Certain landforms also are related to the deposition of till by the glacier ice. *Moraine* is the general term that is used and comes from the French who applied it to ridges of rock and earth around the margins of glaciers in the Alps. Two major kinds of moraines are end moraine and ground moraine. An *end moraine* is a ridgelike accumulation of till with associated stratified drift built along the margin of a glacier. The ridges are aligned in a systematic pattern on the landscape. Commonly, many ridges may be parallel or subparallel and extend across country for great distances, forming an end-moraine system or belt. The Bemis, Altamont, Humboldt, and Algona moraines in Iowa are such systems (pl. 1; Fig. 2.9).

Ground moraine is distinguished from end moraine in that the topographic highs and lows of the landscape are not aligned in a systematic pattern. These highs and lows are a *swell-and-swale* topography. Instead, the pattern may be random or even haphazard. Till and stratified drift may also be associated in ground moraine. Ground moraine is usually behind or in back of an end moraine in the direction from which the glacier ice advanced.

The association of end and ground moraine is recognized technically in the Pleistocene geology trade in the *morphostratigraphic unit* (Frye and Willman, 1960). This is a unit composed of a body of till and stratified drift that is identified primarily from the surface form that it displays. Each unit contains the end moraine and ground moraine, and the continuation of the body beneath the surface where it can be recognized. On the Cary drift in Iowa, the Bemis, Altamont, Humboldt, and Algona systems are recognizable morphostratigraphic units. The moraine components can be identified with ease and can be mapped readily (pl. 1; Fig. 2.9).

The terminal margin of an end-moraine system marks the limit to which glacier ice covered a region. At the maximum advance of ice that formed the Bemis moraine in Iowa, about 12,300 square miles were covered by ice in the north central part of the state. At the Altamont maximum, about 7,030 square miles were covered. At the Humboldt maximum, about 3,170 square miles were under ice, and at the Algona maximum, about 1,650 square miles were covered. These areas in percent of the total area of the Des Moines lobe are Bemis—100, Altamont—57, Humboldt—26, and Algona—13. From the maximum coverage during Bemis time the glacier successively wasted to about a half of its previous size to Altamont, Humboldt, and Algona times. Within the glacial history, these systems represent retreatal phases of the ice margin, with the ice pausing long enough to build the end moraines.

During construction of each system, the drift was successively built up like layers of a cake. A few miles south of New Providence in extreme southern Hardin County the surface of the Bemis end moraine stands 84 feet above the surface of the loess-mantled pre-Cary land surface 2 miles to the east (pl. 1). Seven miles south of Hampton in southern Franklin County the Bemis-moraine surface stands 62 feet higher than the pre-Cary land surface 1½ miles to the east. In Boone County the Altamont-moraine surface stands 33 feet higher than the Bemis-moraine surface 1 to 2½ miles south of Pilot Mound. Along State Highway 60 the difference in relief is 50 feet, but 4 miles to the east the difference between the Altamont and Bemis is 100 feet.

Within an end-moraine system, detail of aligned ridges may be quite complex (Fig. 2.10). In the Bemis system at the southern part of the Des Moines lobe, the ridges are numerous, arcuate, and convex toward the margin of the lobe.

FIG. 2.10. Detail of aligned ridges within end-moraine system. (A) Mapped crest lines of ridges across four counties in central Iowa. See plate 1 for location on Des Moines lobe. Altamont moraine extends across Greene and Boone counties in northern townships. Note looped patterns and truncation of one pattern by another. (B) Airphoto mosaic of area near Nevada in Story County. Ridges show as light pattern, and swales show as dark pattern. At least 19 ridges are in section 10. Scale is 1 mile. (C) Soil map of same area as B. Clarion soils are on ridges and Webster soils are in depressions. Note alignment of soils. Scale is 1 mile. (B and C are from Gwynne and Simonson, 1942.)

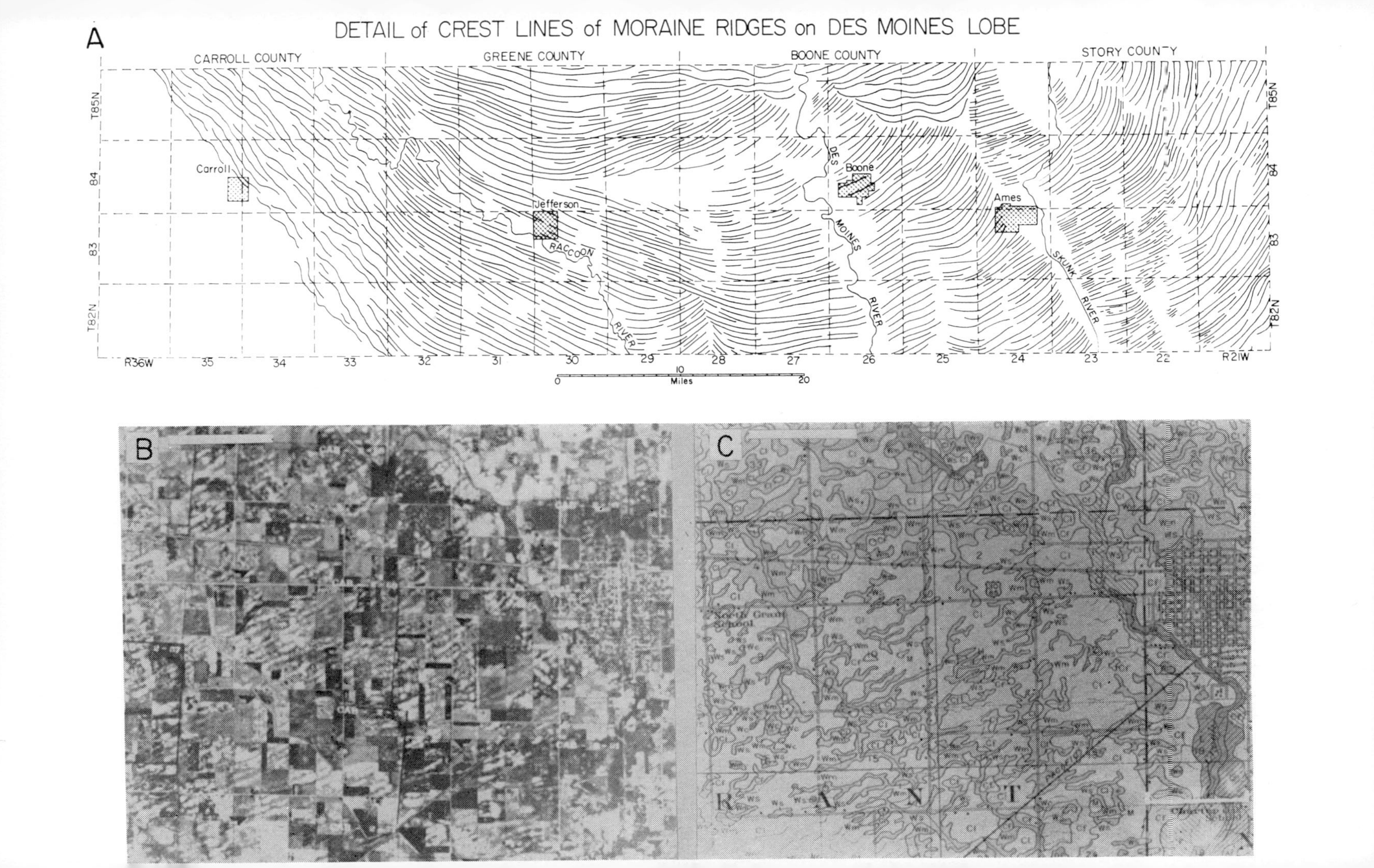
A
DETAIL of CREST LINES of MORAINE RIDGES on DES MOINES LOBE
CARROLL COUNTY
GREENE COUNTY
BOONE COUNTY
STORY COUNTY
T85N
84
83
T82N
R36W
35
34
33
32
31
30
29
28
27
26
25
24
23
22
R21W
Carroll
Jefferson
Boone
Ames
RACCOON
RIVER
DES
MOINES
RIVER
SKUNK
RIVER
0
10
20
Miles
B
C

The patterns loop and cut across each other. In section 10, T. 83N., R. 23 W., Story County, at least 19 ridges are parallel and trend southwest to northeast. Each of these ridges presumably marked the margin of the ice as it retreated to its temporary stand at the Altamont moraine. This intricate pattern where relief is only 5 to 10 feet has been termed *minor moraines* (Gwynne, 1942, 1951) or *washboard moraines* (Lawrence and Elson, 1953).

The Des Moines drift lobe, then, can be pictured at three levels of magnitude: (1) The lobe itself occupies 12,300 square miles. This is the broad picture of Cary glaciation in Iowa. (2) The lobe is divisible into four major end- and ground-moraine systems. Each of these systems marks a pause and a retreatal phase in the general shrinkage of the glacier. (3) The extreme detail of the ridges within the systems shows the positions of the margin of the ice within each phase of retreat. The time element between ridges has been proposed as annual or cyclic, but to date this has not been proved.

Dating the Drift

Numerous radiocarbon dates have been determined for various parts of the drift of the Des Moines lobe (Table 2.7). They can be related to the end- and ground-moraine systems and to the vertical stratigraphy of the drift within a system. However, certain discrepancies arise in the overall pattern of the chronology of dates. One problem involved is that the samples were analyzed by different laboratories. A second problem is that each laboratory used different methods to analyze the samples. Some of the dates were determined using the original carbon black method. Some were reported on the basis of the acetylene gas counting technique, and some were reported where the carbon dioxide gas method was used. The techniques of measurement of the radiocarbon content of organic matter have been considerably improved since the original solid carbon method was used (Libby, 1965). Only slight

TABLE 2.7. RADIOCARBON DATES OF THE CARY DRIFT IN IOWA

Sample*	Date in years before present	Location	Notes
Algona moraine			
W-626	12,970 ± 250	Britt, Hancock County	Larch wood
W-625	13,030 ± 250	Britt, Hancock County	Peat
Humboldt moraine			
C-912†	12,120 ± 530	Lizard Creek, Webster County	Hemlock wood
C-913	13,300 ± 900	Lizard Creek, Webster County	Hemlock wood
Altamont moraine			
I-1414	14,500 ± 340	McCulloch bog, Hancock Co.	Basal muck in bog
Bemis moraine			
C-596	11,952 ± 500	Cook Quarry, Story County	Hemlock wood
C-563	12,200 ± 500	Cook Quarry, Story County	Hemlock wood
I-1015	13,775 ± 300	Colo bog, Story County	Basal muck in bog
W-513	13,820 ± 400	Scranton No. 1, Greene County	Spruce wood
I-1268	13,900 ± 400	Stratford, Hamilton County	Spruce wood
W-517	13,910 ± 400	Scranton No. 2, Greene County	Spruce wood
C-664	14,042 ± 1,000	Cook Quarry, Story County	Hemlock wood
I-1402	14,200 ± 500	Nevada, Story County	Spruce wood
W-512	14,470 ± 400	Scranton No. 1, Greene County	Fir, hemlock, larch, and spruce wood
W-153	14,700 ± 400	Clear Creek, Story County	Hemlock wood

* Sample numbers are C for University of Chicago; I for Isotopes, Inc.; and W for U.S. Geological Survey, Washington, D.C.

† All Chicago dates are by original carbon black method. Isotope dates are by carbon dioxide gas method. U.S. Geological Survey dates are by acetylene gas method.

variation in dates may appear to be significant discrepancies in the chronological system, particularly when the total time span is only 1,000 years. However, careful arrangement of the values in their natural setting and within the framework of methods of analysis results in an orderliness of the system.

The base of the Cary till in the southern part of the Des Moines lobe and through the latitude of Ames is reasonably established at 14,000 years. Seven dates from localities across the lobe (pl. 1) range from 13,820 years (W-513) to 14,700 years (W-153; see Table 2.7). One solid carbon date, C-664, is in agreement, but two values, C-596 at 11,952 years and C-563 at 12,200 years, are not. These last two samples are internally consistent in value and were collected from the same stratigraphic horizon (see Radiocarbon Catalog, Chapter 6). They are 2,000 years younger than sample C-664, which is only slightly lower in the stratigraphic section. They are discrepant and questionable values.

A further check on the validity of the younger dates is the internal consistence of basal till dates at Ames and Nevada with a date from the base of a peat bog on top of the till at Colo. All sites are in Story County (pl. 1). The basal dates are 14,700 years (W-153) west of Ames; 14,042 years (C-664) northeast of Ames; and 14,200 years (I-1402) at Nevada. The date on top of the till at Colo is 13,775 years (I-1015; see Table 2.7). The dates of about 12,000 years of solid carbon dating that are 2,000 years too young are questionable.

Not only is the base of the Cary till easily identified in the field, which permits accurate radiocarbon sampling (Fig. 2.11), but major end-moraine systems or, from the technical point of view, the morphostratigraphic units, are readily separated from adjacent units and dated. To understand this problem, a coniferous forest buried under outwash near Britt in Hancock County may be examined. For orientation refer to plate 1, Figure 2.9 and Figure 2.12. In this area outwash spreads from the terminus of the Algona end moraine and buries 59 square miles of Altamont ground moraine. Some of this outwash spills through gaps in the Altamont end moraine, mixes with Altamont outwash, and buries 27 square miles of Bemis ground moraine. These areas of stratified drift are a few of the true outwash plains in Iowa. Two and a half miles north of Britt a dug drainage ditch exposes Algona outwash that buried a coniferous forest that was growing on the Altamont ground moraine (Figs. 2.11, 2.12). A large tree stump rooted in place in peat interbedded in the outwash is 12,970 years old (W-626), and the peat is 13,030 years old (W-625; see

FIG. 2.11. (A) Stratford section, Hamilton County, showing Cary till (1) on radiocarbon horizon and (2) over pre-Cary silts, sands, and till. (B) Coniferous forest buried by Algona outwash exposed along drainage ditch near Britt, Hancock County. Note numerous stumps on both banks. (C) Detail of stump of larch tree of B.

A
1
2
B
C

FIG. 2.12. Site of the Britt forest bed relative to the Algona and Altamont moraines and their associated outwash plains.

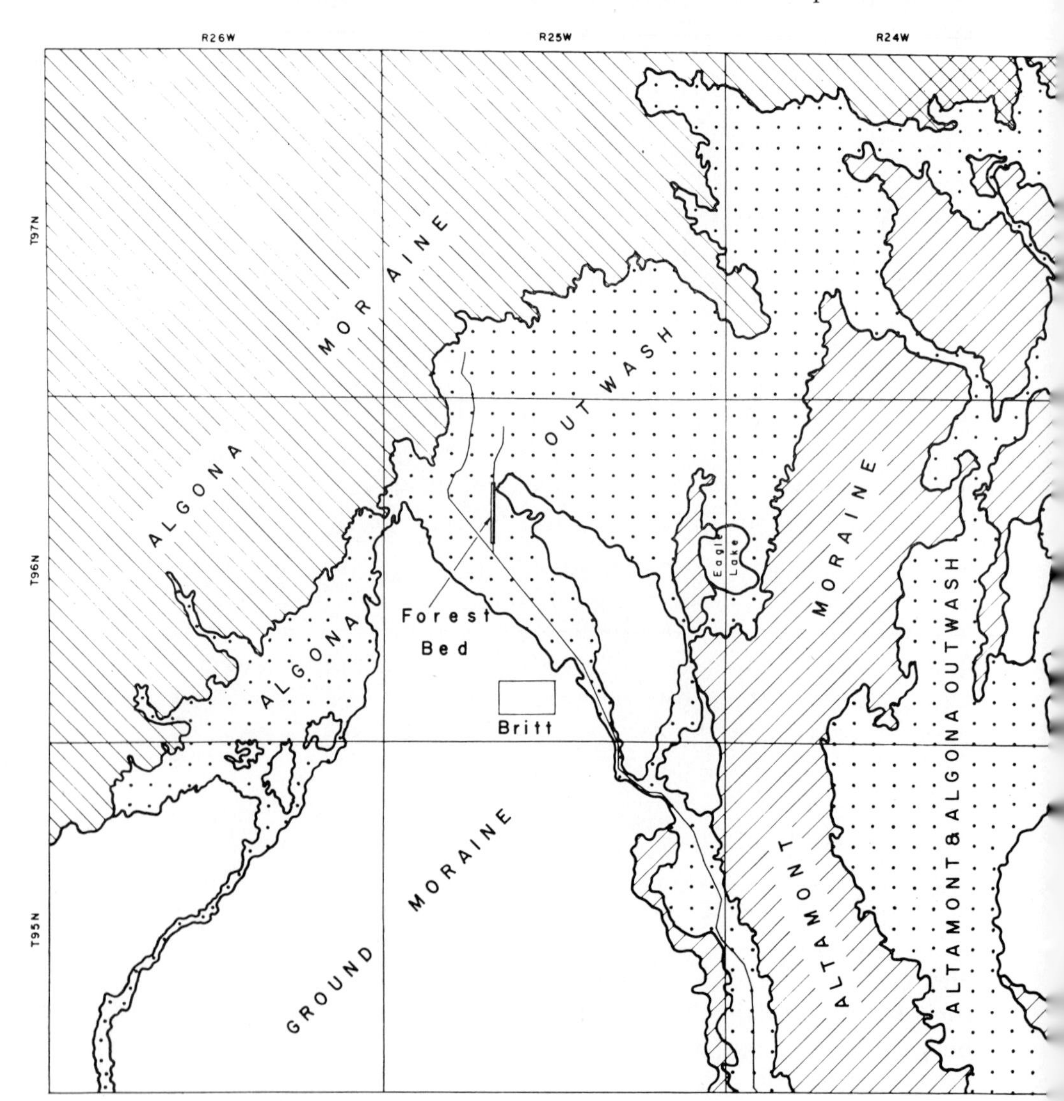

Table 2.7). These dates are almost identical and for all practical purposes, the maximum age of the Algona system is 13,000 years.

What does this mean in terms of glacial history of the Des Moines drift lobe? Glacier ice deposited basal till at the latitude of Ames 14,000 years ago and had to advance to the terminal end moraine at Des Moines 32 miles to the south. The drift surface at Colo, latitude of Ames but only 4 miles from the eastern lobe margin, had to be uncovered by the ice at 13,775 years ago (I-1015). The margin of the ice mass had to waste and shrink so that the terminus was at the Algona position 13,000 years ago. This is a distance of 109 miles north of Des Moines. During 775 years all of the complex washboard moraine of the Bemis-moraine system (Fig. 2.10), the Altamont end-moraine system and ground moraine, and Humboldt end and ground moraine were constructed. All of the features of the entire lobe had to be formed in 1,000 years. This is fast geologically speaking and in geologic thinking.

Do other dates given in Table 2.7 fit in the general chronology? The basal muck in the McCulloch bog on the Altamont moraine in Hancock County is 14,500 years old, which equates with the dates at the base of the Bemis moraine. Perhaps the sample is contaminated by older carbon or perhaps this particular depression existed through Bemis and Altamont time, unhindered by ice so that sediment could be trapped. Neither explanation is satisfactory.

Do dates of the Humboldt moraine, which are 12,120 ± 530 years (C-912) and 13,300 ± 900 years (C-913; see Table 2.7), fit in the chronology? The solid carbon method enters again. These samples were two pieces of hemlock wood only a few feet apart and along the same bedding plane in sands and gravels in till. Somehow one is 1,000 years younger than the other. Bets should be placed on the older date that fits between 13,000 and 13,775 years.

All of this brings us back to the term Cary, and why Cary? On the basis of regional mapping and correlation, a part of the drift of the Des Moines lobe was named Cary and a part was named Mankato (Ruhe, 1952a, b). These two drifts were recognized at that time as the third and fourth substages of the Wisconsin glacial stage in the Midwest. This classification was made prior to having radiocarbon dates. Then the dates poured in from many places, and those on the Des Moines lobe fit into the group of dates of 12,000 to 14,000 years that typified Cary throughout the Midwest (Flint and Rubin, 1955). The Cary of the Des Moines lobe was in business, but the Mankato had to be given the business (Ruhe and Scholtes, 1959). All of the Des Moines lobe-moraine systems in Iowa were in the older part of the age range. Now, the name Cary has been changed in many places, including the type area near Chicago, Illinois (Frye and Willman, 1960). But in Iowa we remain old-fashioned and stubborn; the name remains here.

CHAPTER 3

BENEATH THE WISCONSIN LOESS

IN THE CASE OF THE CARY DRIFT, WE moved upward stratigraphically from our key bed, the Wisconsin loess. Using the same key, we can now move downward and study the deposits and land surfaces beneath the loess in the parts of Iowa beyond the limits of the Des Moines lobe. Our procedure will be to return to southwestern Iowa, cross the state eastward to southeastern Iowa, then move northward to the northeastern part of the state. Next, moving westward, the Cary drift lobe will be jumped so that the relations in the northwest part of the state may be studied.

Beneath the Wisconsin loess in these areas are features which are analogous to things that can be observed and measured on the present land surface. There are buried landforms, landscapes, and soils (Fig. 2.1). If geomorphology is the study of landforms, and pedology is the study of soils on the landforms, do the buried landforms and soils fall within the domains of these sciences? Yes, but there must be qualification. Usually geomorphology and pedology pertain to the present land surface. Where land surfaces and soils are buried, they may be appreciated within frameworks of *paleogeomorphology* and *paleopedology*. These are no more than reasonable applications of the principle of uniformitarianism, that the present is the key to the past. This reasoning implies

(Thornbury, 1954): (1) The same processes that operate at present also operated in the past, although not necessarily with the same intensity. (2) Geomorphic [and pedologic] processes leave their imprint on landforms, and each geomorphic process develops a characteristic assemblage of landforms. (3) But, complexity of geomorphic [and pedologic] evolution is more common than simplicity.

PALEOGEOMORPHOLOGY AND PALEOPEDOLOGY

The prefix *paleo* is from the Greek *palaio* meaning old or ancient. How old is ancient? We might say as remote as necessary to represent the past. In some cases yesterday may be old enough, but in other cases thousands of years may be required. Thus, in paleogeomorphology, we may recognize paleo-landscapes, and in paleopedology we may recognize paleosols (*sol* being from the Latin, *solum,* or soil). Combining all of these terms, a paleosol is a soil that formed on a paleo-landscape.

There are three basic kinds of paleo-landscapes and paleosols. They are buried, exhumed, and relict (Ruhe, 1965). *Buried* soils formed on preexisting landscapes and were subsequently covered by younger sediment or rock. These soils and land surfaces crop out in natural or man-made excavations, such as stream or road cuts, and on hillslopes. They occur beneath many different kinds of deposits, including loess and glacial till in the Midwest, alluvial-fan gravels in the southwestern deserts, and even beneath lava flows in Hawaii (Figs. 1.2, 1.3).

Exhumed soils are those that were buried but have been reexposed on the present land surface by erosion of the covering mantle. The old land surface is also exhumed. These features may have wide extent and may give rise to anomalous relations on the land surface. They may be side-by-side with other soils and landforms that are more in harmony with the present environment.

Relict soils formed on preexisting landscapes but were never buried by younger sediments and, of course, are not exhumed. These soils date from the initiation of the original land surface. A simple example of relict soils and relict landscape is three terraces along a stream valley. The land surface of the highest terrace and the soils on it began to form prior to the formation of the lower two terraces. Consequently, the high terrace and its soils may have some relict features of the past. The other terraces and their soils did not experience that part of the valley history represented by the high terrace. On the other hand, the high terrace did experience the episodes of valley history represented by the lower terraces.

This brief background will provide a means of communicating about the things beneath the Wisconsin loess in Iowa.

SANGAMON SURFACES AND UNDERLYING DEPOSITS

Just beneath the Wisconsin loess in southwestern Iowa between the Missouri River Valley and eastward to a limit that extends from Carroll County southward to western Cass County and middle Page County is a well-developed paleosol in a lower loess (pl. 1). The paleosol is the well-known Sangamon soil that marks the third interglacial episode of the Pleistocene throughout the Midwest and westward on the Great Plains. The lower-lying loess is the Loveland, named by Shimek (1909) after the village of Loveland in the northwestern corner of Pottawattamie County. Here is another Pleistocene type locality of world reknown that is located in Iowa. The Loveland loess is recognized in the Missouri Valley region, on the Great Plains, and down the Mississippi Valley to the Gulf of Mexico (Thorp and Smith, 1952). In order to understand the Sangamon paleosol, it is necessary first to study the Loveland loess which is the sediment in which the Sangamon soil is formed.

Loveland Loess

The sedimentological system of the Loveland loess is very similar to that of the Wisconsin loess in southwestern Iowa. To demonstrate this fact, we can return to the loess traverse along the Rock Island Railroad from Bentley in Pottawattamie County to Atlantic in Cass County. Properties of the Loveland loess were measured at the same places and directly beneath the Wisconsin loess in the railroad cuts (Fig. 2.1). Thus, a direct comparison can be made between the two loesses.

First, the Loveland loess was also deposited on a preexisting surface that had topographic highs and lows. This preexisting land surface represents the Yarmouth or the second interglacial episode, because the Loveland loess is Illinoian in age (Leighton and Willman, 1950). The preexisting topography had ridges that rose like a staircase from one main valley to an interstream divide and then descended a staircase to the next main valley to the east. The minimum elevation at which this loess was deposited is 1,180 feet, and the maximum elevation is 1,300 feet. The difference is 120 feet. In discussing the Wisconsin loess, we noted that the loess crossed the staircase of ridges along this traverse through an elevation range of 218 feet with the minimum and maximum elevations at the base of the loess being 1,100 and 1,318 feet, respectively. These elevations at the base of the Loveland loess are 1,180 to 1,300 feet. Both loess systems in this regard are alike in kind, but the absolute values differ between systems.

Particle size also is systematically distributed along the traverse as the Loveland loess thins (Fig. 3.1). The distributions vector from west to east even though the loess was deposited on preexisting rough topography. The loess is 313 inches

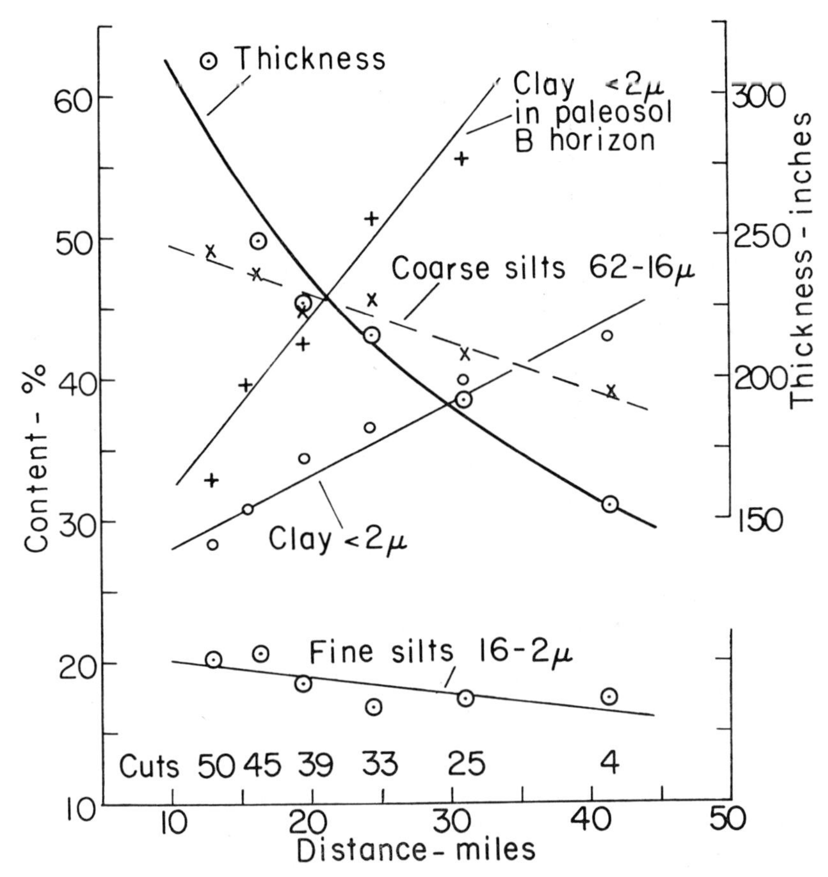

FIG. 3.1. Relations of thickness and coarse silt, fine silt, and clay content of Loveland loess to distance along traverse from Bentley to Atlantic in southwestern Iowa. Missouri River Valley source is 13 miles west (left on diagram) of cut 50. Amount of clay in B horizon of Sangamon paleosol increases eastward from cut 50 to cut 25.

thick at cut 50 and is 155 inches thick at cut 4. Thinning progresses from a distance of 13 miles from the source at cut 50 to 16.3 miles at cut 45, 19.6 miles at cut 39, 24.4 miles at cut 33, 31 miles at cut 25, and 41.4 miles at cut 4. Other loess thickness values are 248, 227, 215, and 192 inches at cuts 45,

39, 33, and 25, respectively. How does this look mathematically? The loess thins as expressed by:

$$Y = 1/(2.13 \times 10^{-3} + 1.04 \times 10^{-4} X)$$

where Y is the thickness in inches and X is the distance in miles. This rather complicated-looking equation is really the simple old friend, the hyperbola, that has the general expression, $Y = 1/a + bX$. We will also recall that the Wisconsin loess thinned in a like manner. The only dissimilarity is that the constants in the two equations differ. The reason that they do is due to absolute differences in loess thickness at the same site. For example, at cut 50 the Loveland loess thickness is 313 inches, but Wisconsin loess thickness is 453 inches, and so on for the other sites.

Course silt content (Y) in the Loveland loess decreases progressively with distance (X) from 49 percent at cut 50 to 38.9 percent at cut 4. This relation is expressed as:

$$Y = 52.885 - 0.345X$$

Note this similarity to the Wisconsin loess at the same sites, but this relation is linear, whereas in the Wisconsin loess the relation is slightly curvilinear.

Fine silt content (Y) in the Loveland loess decreases progressively with distance (X) from 20.1 percent at cut 50 to 17.4 percent in cut 4. This relation is expressed as:

$$Y = 21.239 - 0.113X$$

but differs from that of the Wisconsin loess where fine silt content increased slightly in a curvilinear relation. Why is there a difference? The Loveland loess is altered more and is progressively more thoroughly weathered to the east. Fine silt has decomposed to form clay, but in the Wisconsin loess this is not so.

Note the amount of clay (Y) in the Loveland loess which progressively increases with distance (X) from 28.2 percent in cut 50 to 42.8 percent in cut 4. This relation is expressed by:

$$Y = 22.944 + 0.511X$$

and is the same kind of relation as in the Wisconsin loess. However, in the latter, clay content only increases slightly from 19 to 23 percent. Consequently, at the same distances from the Missouri River Valley source, the Loveland loess is appreciably more weathered than the Wisconsin loess.

Regardless of this alteration, the remarkable thing about this business is that two eolian sediment systems should be so similar even though they are so widely separated in time. The Loveland system formed during the Illinoian or third glacial episode and the Wisconsin loess formed during a part of the Wisconsin or fourth glacial episode. The absolute age of the Loveland loess cannot be fixed with certainty. It is older than 23,900 (I-1420) and 24,500 years (W-141), which are the dates at the base of the Wisconsin loess which buries the Loveland (Table 2.3). It is also older than 37,600 ± 1,500 years (W-880), which is the terrace alluvium along the Boyer River at Logan, Iowa. This alluvium is a valley fill inset below the Loveland loess that is buried on the adjacent uplands.

The Loveland loess thins to oblivion in Audubon, Cass, Montgomery, and Page counties (pl. 1). To the east it cannot be recognized between the Wisconsin loess and the paleosol on the Kansan till. To the west it is recognizable and measurable beneath these two features. Consequently, in relative dating the Loveland is younger than the paleosol (Yarmouth) on the Kansan till and older than the Wisconsin loess which buries it. This makes the Loveland Illinoian in age in southwestern Iowa.

Sangamon Surface and Paleosol

The Sangamon paleosol on Loveland loess differs distinctly from soils formed on the overlying Wisconsin loess at the same locality. The Sangamon paleosol is redder and has somewhat "stronger development." The term soil development is a catch-all which implies that certain soil properties are shown better in one soil than in another. For example, the Sangamon soil on Loveland loess has been recognized in the Mississippi River Valley in the states of Tennessee and Mississippi (Wascher *et al.,* 1947). In Obion County, Tennessee, contrasts of properties of the Sangamon soil versus the land surface soil on Wisconsin loess are respectively: (1) light brown versus yellowish-brown A horizons, (2) yellowish-red versus brown B horizons, (3) clay content ($< 2\mu$) of 35 percent versus 27 percent in the B horizons, and (4) solum thickness of 36 versus 27 inches. Consequently, the Sangamon paleosol is interpreted as a "more strongly developed" soil.

Returning to our now overworked loess traverse between Bentley and Atlantic in southwestern Iowa, the Sangamon soil can be studied in ridge after ridge in the railroad cuts. Comparison can be made directly above to the soil on the land surface on Wisconsin loess. At all places the Sangamon paleosol is much more developed than the land surface soil (Fig. 2.1). The paleosol B horizons are reddish brown and when observed from a distance, extend through the cut as a reddish-colored layer. The surface soil B horizons generally appear as a brown or yellowish-brown layer.

From east to west along the traverse, the amount of clay in the paleosol B horizons increases very systematically (Fig. 3.2). It is 32.9 percent at cut 50 and 55 percent at cut 25. The relation of clay content (Y) to distance (X) is expressible mathematically as:

$$Y = 19.72 + 1.23X$$

A statistical test of this equation, standard error of estimate, is 1.89 percent. Hence, at any point along the traverse, the clay content of the paleosol B horizon may be estimated reasonably closely if the distance from the source is known.

The factor *distance* cannot be responsible for the change

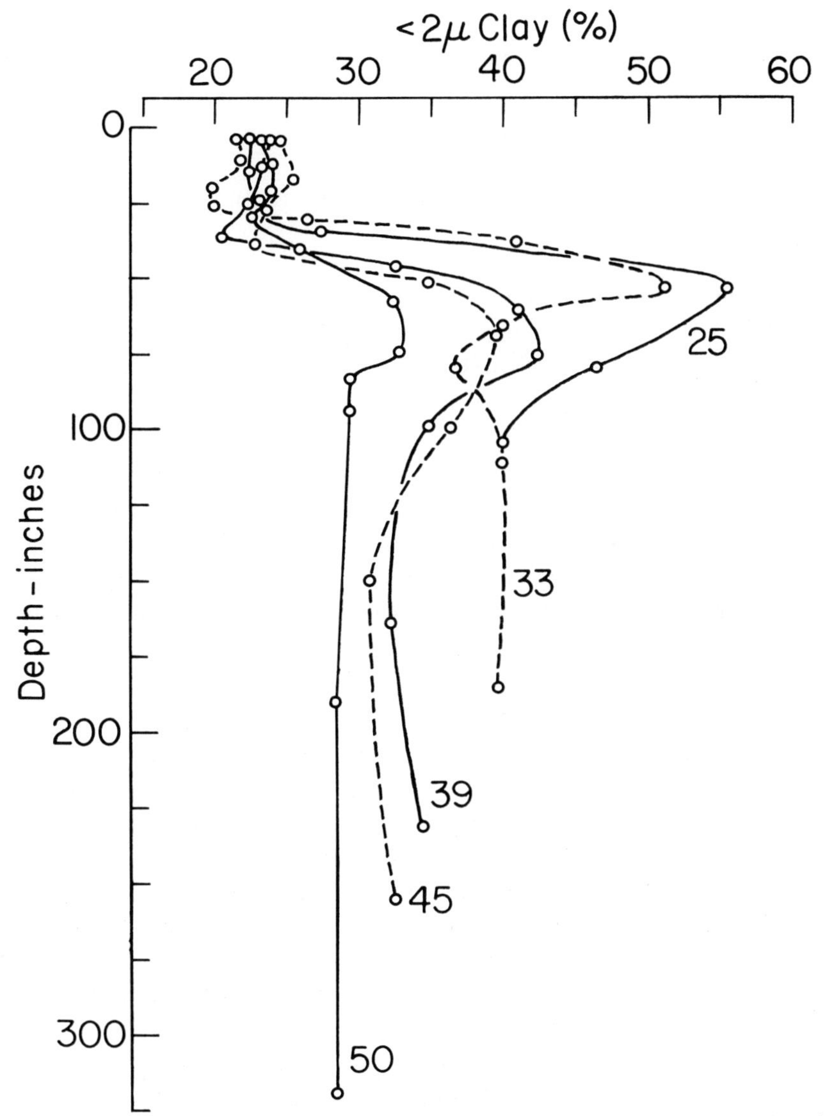

FIG. 3.2. Clay content of the B horizons of Sangamon paleosol from west (cut 50) to east (cut 25) along Bentley to Atlantic traverse in southwestern Iowa.

in B horizon property. But we noted previously that the Loveland loess thins to the east, that the loess is deposited on a more impermeable paleosol, and that the particle sizes of the loess systematically decrease to the east. The relation of clay content of B horizon (Y) to thickness of the loess (X) is very systematic as expressed by:

$$Y = 1/(-1.09 \times 10^{-3} + 1.03 \times 10^{-4}X)$$

Here is our previously noted curve of the hyperbola which indicates that the rate of increase of clay of the B horizons progressively changes per equal unit of distance along the direction of loess distribution. Why? Let us hold this question in abeyance until we examine the system of soils on the present land surface in relation to Wisconsin loess in the same region.

However, we can examine one additional property for comparison purposes. In the Sangamon paleosols the thickness of the solum (recall that solum is the A and B horizons of the soil) ranges from 55 to 80 inches with an average of 67 inches. In the land surface soils, thickness ranges from 24 to 63 inches with an average of 47 inches.

What does all of this mean? The soil formed during the third interglacial episode in Iowa is more deeply and strongly developed than surface soils that formed on the land surface since a point in time during the last or Wisconsin glaciation. One reason is that the time of the third interglacial or Sangamon was relatively longer than since that point in time during the Wisconsin. This last increment of time is 14,000 years, so the Sangamon must have been longer. Factors other than time may enter the picture and cause strong soil development. We will look at these factors later in discussing environment.

The Sangamon paleosols formed more than 24,000 years ago, as the base of the Wisconsin loess that overlies them is 23,900 years (I-1420) and 24,500 years (W-141) old. If the previous calculation of 14,000 years is added to these values, Sangamon time in Iowa was 38,000 years ago and older. If the 14,000-year value is added to the age of terrace alluvium at 37,600 years (W-880) at Logan, Iowa, a date in Sangamon time would be about 52,000 years ago. Undoubtedly, it is much older than that.

Sangamon paleosols are present also on the Illinoian glacial drift in southeastern Iowa (pl. 1). Many of these paleosols formed on the till have been known as "Illinoian gumbotil" (Kay and Graham, 1943). Some of the characteristics of the gumbotil were introduced previously (see Background), but we shall take a closer look at their pedological nature in a following section.

Late Sangamon Surface and Paleosol

We know that Sangamon time in Iowa had considerable duration because two distinct episodes can be recognized within it. First, there was the formation of the Sangamon paleosol on Loveland, and second, a widespread erosion surface formed, then stabilized, and another episode of soil formation followed. The second part is the Late Sangamon, and its surface and soil also are buried beneath the basal soil of the Wisconsin loess.

How is the Late Sangamon surface and its paleosol separated from the Sangamon surface and its paleosol? The separation involves basic principles of the geologic unconformity in the stratigraphic layering of sediments in southwestern Iowa. An *unconformity* is a surface of erosion or nondeposition, usually the former, that separates younger from older strata. The Late Sangamon surface classically illustrates this principle in the Rock Island Railroad cuts on opposite sides of the West Nishnabotna River near Hancock, Pottawattamie County. Cut 33 is a divide to the west of the river, and cut 25 is a divide to the east of the river (Fig. 3.3). Stepping down the staircase of ridges westward from cut 25 at the east divide to cut 31 on the east valley slope, progressively older Pleistocene deposits are beneath the basal soil of the Wisconsin loess.

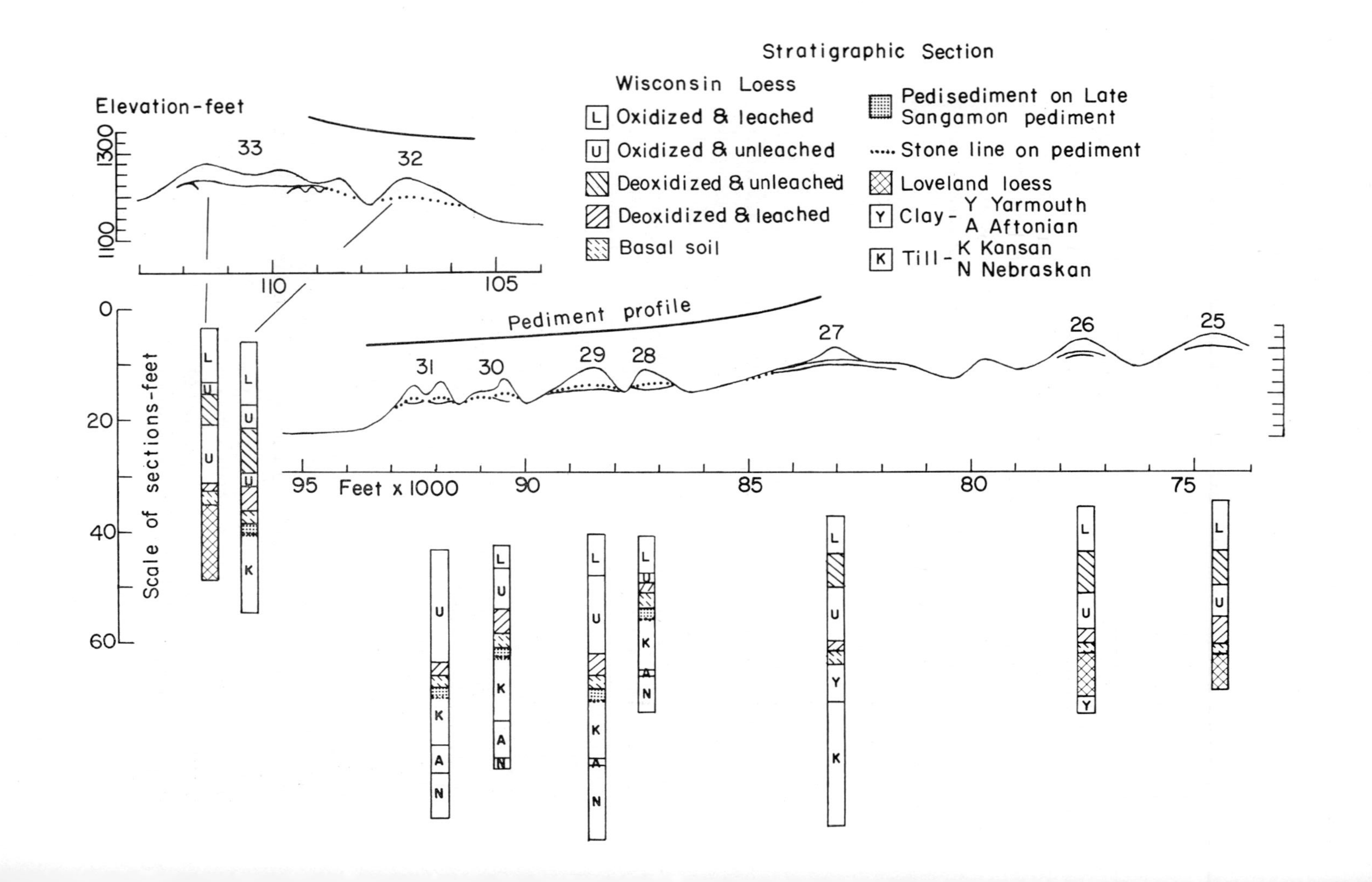
Stratigraphic Section
Wisconsin Loess
L Oxidized & leached
U Oxidized & unleached
Deoxidized & unleached
Deoxidized & leached
Basal soil
Pedisediment on Late Sangamon pediment
Stone line on pediment
Loveland loess
Y Clay - Y Yarmouth A Aftonian
K Till - K Kansan N Nebraskan
Elevation-feet
1300
1100
33
32
110
105
Pediment profile
31
30
29
28
27
26
25
Scale of sections-feet
0
20
40
60
95
Feet x 1000
90
85
80
75

In cuts 25 and 26 the Sangamon paleosol on Loveland loess is beneath the basal soil. In cut 27 the Loveland loess is absent and a Yarmouth clay is in place. In cuts 28 and 29 the Loveland loess and Yarmouth clay are absent and a stone line surface on Kansan till is present. In cuts 30 and 31, an Aftonian clay is beveled, and a stone line surface on Nebraskan till is present. The basal soil of Wisconsin loess overlies all of these features. At cut 25 the normal relation exists with basal Wisconsin soil on Sangamon surface or Wisconsin on Sangamon. But, in cut 31 basal Wisconsin soil overlies Nebraskan till. Here the Aftonian, Kansan, Yarmouth, and Illinoian are missing. This hiatus has considerable magnitude and includes the stratigraphic record of major parts of the Pleistocene in Iowa. A *hiatus* is a gap in a rock sequence represented by geologic deposits which normally would be present, but are missing due to the fact that they were never deposited or were eroded prior to deposition of the immediately overlying beds.

On the west side of the Nishnabotna Valley (Fig. 3.3) the basal soil of the Wisconsin loess overlies the Sangamon paleosol on Loveland loess in cut 33, and then is over the stone line on Kansan till in the east part of the cut and also in cut 32.

The most distinct field evidence for the erosion surface is the stone line that crosses through the cuts at the top of the tills. A *stone line* is a concentration of coarse rock fragments in soils or sediments that may be one stone thick or more than one stone thick, that overlies material from which the stones were concentrated, and which is generally overlain by variable thicknesses of finer textured sediment (Ruhe, 1959; Fig. 3.4). The stone line surface extends 1⅔ miles east of the east valley wall of the Nishnabotna and ¾ mile west of the west valley wall (Fig. 3.3).

Applying relative dating principles, the age of the surface can be determined. The youngest deposit that is cut is the

FIG. 3.3. Stratigraphic and geomorphic relations of the Late Sangamon erosion surface on opposite sides of the West Nishnabotna River near Hancock, Iowa.

Loveland loess. The youngest surface that is cut is the Sangamon surface marked by its paleosol. The erosion surface is younger than the Sangamon paleosol on Loveland loess. The erosion surface is overlain by the basal soil of Wisconsin loess and must be older than Wisconsin. The surface must be placed in the niche between the Sangamon soil on Loveland loess and the basal soil of Wisconsin loess. Thus, a Late Sangamon designation must be created.

Associated with the Late Sangamon surface are various kinds of paleosols that are formed partly in tills and other Pleistocene deposits. Some of the soils are also formed in the debris eroded during formation of the erosion surface and

FIG. 3.4. Stone lines. (A) Marking the Late Sangamon erosion surface on Kansan till in railroad cut at the Jesse James monument west of Adair, Iowa. Overlain by Wisconsin loess. (B) As a discontinuity in a Late Sangamon buried soil. Note that part of the B horizon is above and part is below the stone line. Scale in feet.

deposited elsewhere. The properties of the Late Sangamon paleosols can be better understood where comparison can be made to other till-derived soils. Suffice it to mention at this point that the Late Sangamon paleosols have red colors that are similar to the Sangamon paleosol on Loveland loess. They differ distinctly from the brown and yellowish-brown colors of the subsoils on the Wisconsin loess.

These paleosols are older than the radiocarbon ages of 24,000 and 38,000 years that have been introduced previously at several places in our discussion. They have wide occurrence in Iowa and have been identified across the state in the Kansan drift and Illinoian drift areas (pl. 1). They have even been found buried beneath the Des Moines lobe. As one drives the roads in the state and observes a reddish layer with a line of gravel in a fresh road cut or the same features cropping out on a hillslope in a freshly cultivated field, he may be certain that the Late Sangamon surface and paleosol are present.

YARMOUTH-SANGAMON SURFACE AND UNDERLYING DEPOSITS

Across the southern part of Iowa, Wisconsin loess buries prior land surfaces that formed on Kansan drift (pl. 1). This part of the state has been known informally as the Kansan drift region. Kansan till is the deposit on which the previous land surfaces and their associated paleosols and soils formed. The region extends from the eastern limit of the Loveland loess to the western limit of the Illinoian drift.

The topography progressively changes toward the east. Near the Missouri River Valley, slopes are steep, crests of ridges are narrow, and relief from valley floors to ridge crests may be 100 to 120 feet. Eastward slopes progressively flatten, ridge crests progressively widen, and relief progressively decreases to 15 to 20 feet. In south central Iowa the uplands between valleys are broad and extensive and have been termed flat tabular divides (Kay and Apfel, 1929). This is the typical area of the buried Yarmouth-Sangamon surface and its paleosols that have long been called the "Kansan gumbotil." Other former land surfaces also are in this region, including the Late Sangamon surface and others of Wisconsin age. All of these surfaces are formed in a common deposit, Kansan till, so reasonable comparison can be made between paleosols and soils that are on these surfaces. The soil parent material factor is constant.

Where stream incision or artificial excavation has been deep enough, Aftonian paleosols or "Nebraskan gumbotil" and their underlying till may be exposed on hillslopes or in cuts. However, their areas of exposure are small and infrequent, so we shall not concentrate on them. Instead, the major features immediately beneath the loess in the region will be emphasized.

Yarmouth-Sangamon Surface and Paleosols

Under the broad, flat, upland divides and beneath the Wisconsin loess, thick, clayey, considerably weathered paleosols pass from one side of the divide to the other. These paleosols are loosely called by the catch-all "Kansan gumbotil." The term was coined by Kay (1916) who defined *gumbotil* as a gray to dark colored, thoroughly leached, nonlaminated, deoxidized clay, very sticky, and breaking with a starchlike fracture when wet, very hard and tenacious when dry, and which is chiefly the result of weathering of till. Great argument and controversy has arisen recently concerning the nature and origin of gumbotil. This could be resolved readily if all of the things in the catch-all were considered in the context of what they really are, namely, various kinds of paleosols. We shall make such examination.

A second thing is wrong about Kansan gumbotil, and that is the use of the term "Kansan." Weathering of a glacial till is really an interglacial phenomenon. Weathering of Kansan till would have to be at least as young as Yarmouth, the interglacial episode that followed the Kansan glaciation; hence, any stratigraphic naming must use Yarmouth. The same criticism in principle applies to "Illinoian gumbotil" and "Nebraskan gumbotil."

Why have we used the term Yarmouth-Sangamon paleosol for the previously named Kansan gumbotil? The Loveland loess of Illinoian age buries the Yarmouth paleosol in southwestern Iowa. This shows that weathering of the Kansan till had progressed appreciably in Yarmouth time. To the east the Loveland loess thins to where it can no longer be recognized on the Yarmouth paleosol, but at this place and farther to the east, the basal soil of Wisconsin loess directly overlies the buried soil. A hiatus appears again. The Loveland loess and its Sangamon paleosol are missing, but the weathering of the Yarmouth paleosol had to continue through the Illinoian and the Sangamon until burial by Wisconsin loess. Thus, the paleosol in question has to be Yarmouth-Sangamon in order to properly evaluate the magnitude of the hiatus.

Under the tabular divides and beneath the Wisconsin loess (Fig. 3.5) the Yarmouth-Sangamon paleosol is formed in Kansan till and in locally transported sediment on the till surface. This buried surface has an undulating swell and swale surface with a relief of several feet and with distances between swell crests of several hundred feet. On this differing topographic surface are differing kinds of paleosols (Ruhe, 1956). On the crest of a swell the paleosol has an A2 horizon with 32 percent clay that abruptly rests on a B horizon with 58 percent clay (Fig. 3.6, profile D). The solum is 5.7 feet thick. In an adjacent swale a different kind of paleosol (profile F) has a more or less uniform clay content of about 50 percent from top to bottom. The solum thickness is 7.1 feet. A third paleosol (profile G) on the intervening slope between swell and swale has a clay distribution that is similar to parts of the previous two paleosols. Its solum thickness is 7.3 feet. The first paleosol is analogous to a soil on the present surface that

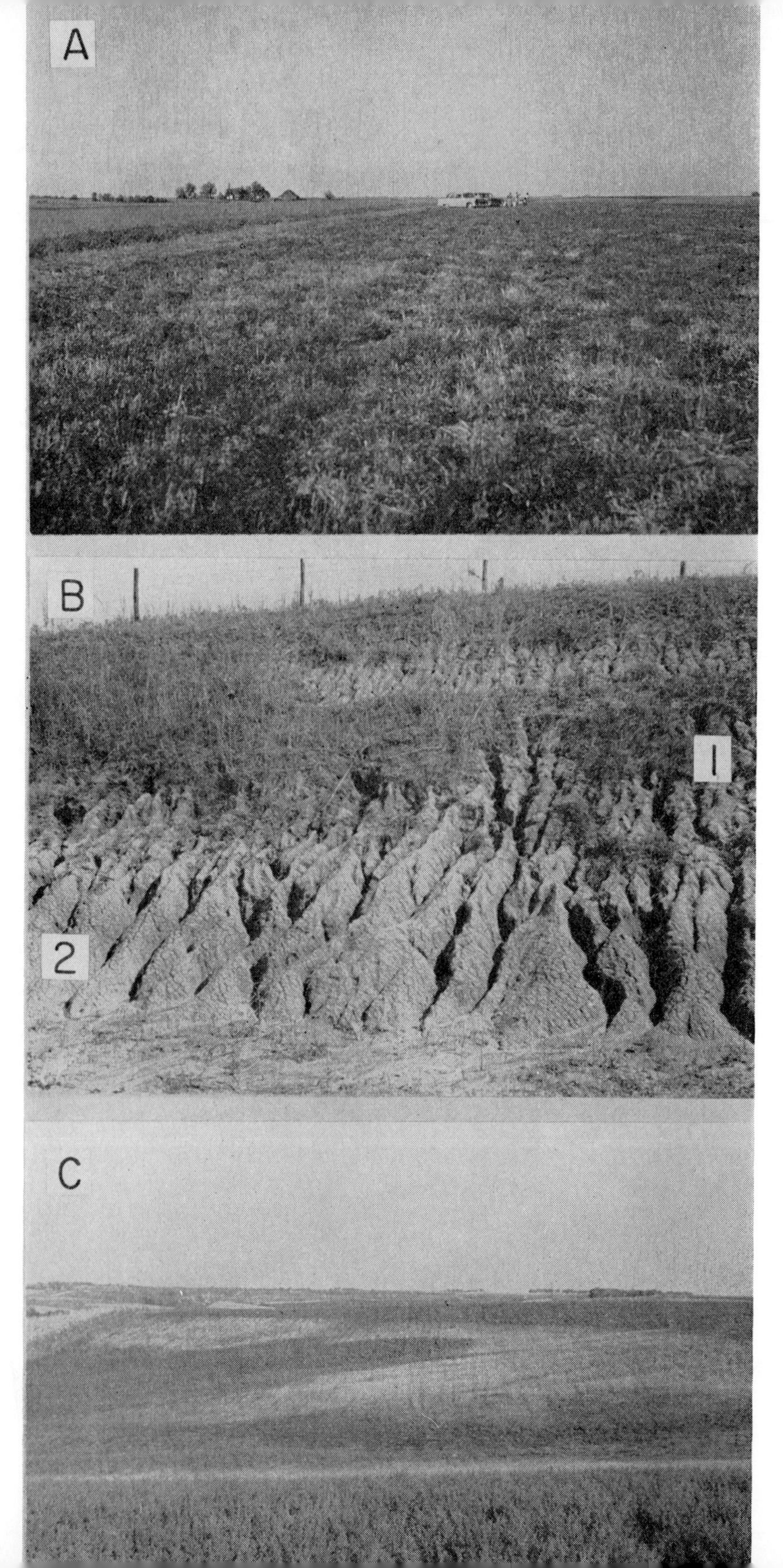
A
B
1
2
C

is called a Planosol or clay-pan soil. The second paleosol is analogous to a Humic-Gley soil or meadow soil. The third paleosol is a combination, an intergrade, or a transition between the other two. On the present land surface of the Cary drift of the Des Moines lobe we can find similar topography with similar kinds of soils in a similar kind of arrangement, but the soils are not so thick, so appreciably weathered, or so strongly developed.

Intensities of weathering can be examined by using weathering ratios of *heavy* and *light minerals* (Ruhe, 1956). These minerals have specific gravities greater or less than that of a heavy liquid such as bromoform whose specific gravity is 2.87 at 20° C. The heavy liquid is used to separate the minerals. Some float and some sink, that is, some are lighter or others are heavier. Experience has shown that heavy minerals such as zircon and tourmaline are more resistant to weathering than an amphibole like hornblende or a pyroxene like hypersthene. In the light minerals, quartz is more resistant than feldspar. The percent by count of resistant and weatherable minerals per size fraction per soil horizon is determined under a microscope, and ratios are calculated. Larger quotients show more resistant minerals than weatherable minerals, and smaller quotients show relatively more weatherable minerals. If the values group closely for samples of the C horizon of the soil, the assumption is made that all parent materials are alike in mineral composition. Larger quotients for B and A horizons than C horizons indicate that less resistant minerals have residually accumulated. In other words, B and A horizons have formed from C horizons.

Applying the weathering index principle to the three Yarmouth-Sangamon paleosols permits an interpretation that

FIG. 3.5. Typical places of the Yarmouth-Sangamon paleosol in southern Iowa. (A) Tabular divide in Adair County. (B) Wisconsin loess (1) burying paleosol (2) on tabular divide. (C) Paleosol cropping out on hillslope at edge of tabular divide. Note poor stand of alfalfa on paleosol (light pattern) with good stands in loess to right and alluvium to left.

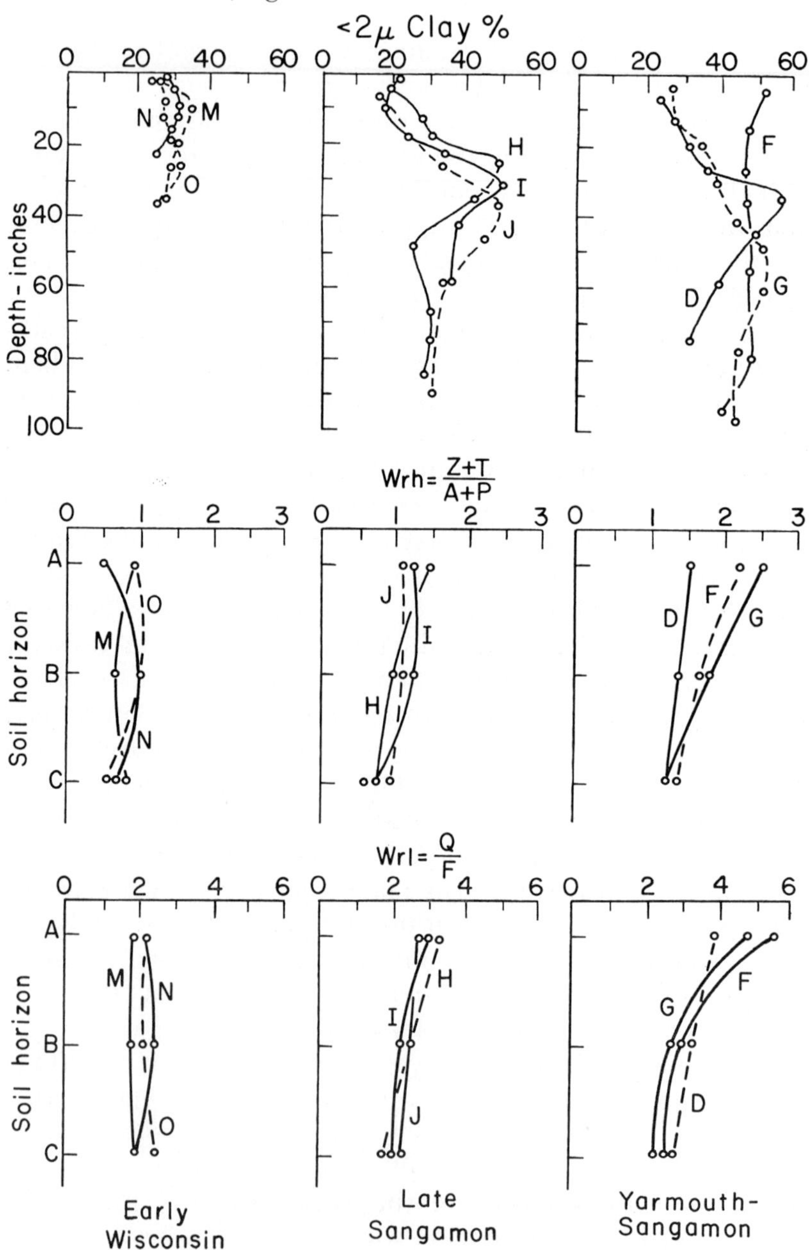

FIG. 3.6. Clay distribution and mineral weathering ratios of Yarmouth-Sangamon and Late Sangamon paleosols and soils on Wisconsin-age surface in Adair County, Iowa. Kansan till is the common parent material. *Wrh* is heavy-mineral ratio with *Z*—zircon, *T*—tourmaline, *A*—amphiboles, and *P*—pyroxenes. *Wrl* is light-mineral ratio with *Q*—quartz and *F*—feldspar. (From Ruhe, 1956.)

the A and B horizons of profile G are more weathered than profile F, which in turn is more weathered than profile D (Fig. 3.6). This is expected. The first paleosol is in a swale; the second is on an intervening slope; and the third is on the crest of a swell. The swale would be more moist than the slope than the swell crest, and, predictably, intensity of weathering would decrease in the order given, and it does.

Present hillslopes bevel the tabular divides in southern Iowa, and on these beveled edges the Yarmouth-Sangamon paleosol crops out on the hillslope (Fig. 3.5). What do we have here? We have an exhumed paleosol. It intervenes between other soils on the landscape and by geomorphic coincidence is brought into association with the other soils. The paleosol is unrelated to the environments that gave rise to the other soils. As shown in Figure 3.5, agronomic problems also are brought into focus. Alfalfa has a poor stand in the clayey, more weathered paleosol, but crop growth is much better on the summit and hillslope soils, above and below the paleosol.

By relative dating techniques we can show that the Yarmouth-Sangamon surface is the oldest paleogeomorphic surface of major extent in Iowa. Its maximum age dates from the beginning of Yarmouth time. Its minimum age is just older than the dates of the basal soil of the Wisconsin loess that buries the surface. The absolute maximum age is not possible to determine in the current state of the art in geochronology. The absolute minimum ages of the surface range from 16,500 years (I-1419A) to 29,000 years (I-1269; see Table 2.3). The lesser value and the many others that occupy the range to 29,000 years also show that the Yarmouth-Sangamon surface was exposed on the tabular divides and subject to weathering well into Wisconsin time. Later increments of Wisconsin loess buried the surface at different times at different places. In fact, the ages of 16,500 to 22,000 years correspond to Tazewell time of the Wisconsin and the ages of 22,000 to 29,000 years equate with Farmdale time of the Wisconsin.

Younger Paleo-Surfaces, Paleosols, and Soils

The land surface in southern Iowa is not so simple as our previous descriptions may have implied. Staircases of ridges and tabular divides are some components that may be set apart. If one starts on a divide and moves along the axis of an interfluve toward a stream valley, he does not descend one long continuous slope. (An *interfluve,* by the way, is the land area between two adjacent streams.) Instead, the long slope is interrupted at several places by distinctly steeper slope gradients. We are back to the staircase again. The steeper gradients that intervene are like the risers of the staircase, whereas the alternate shallower interfluve slopes are like the treads of the staircase. This can be better understood by examining maps (Fig. 3.7). On the topographic map, alternate wide and narrow spacings of contours show the treads and risers along interfluves. For example, from A to P one rises up the valley slope, levels off, rises again, and levels off again at the divide whose elevation is 1,350 feet. From A to F three levels may be determined.

Why are there such levels? The answer can be found in systematically drilling through the Wisconsin loess, measuring each vertical section, and accurately locating each drill hole geographically and in elevation. Three-dimensional reconstruction is then possible. Under the divide and beneath the loess is the Yarmouth-Sangamon surface with the kinds of paleosols that have been previously described (Fig. 3.7). On the first step down and beneath the loess is the Late Sangamon surface with its stone line and reddish-colored paleosols. On the lowest step and beneath the loess is a stone line surface on Kansan till, but there is *no* paleosol as in the uppermost part of the till. This last point is extremely important and will be reintroduced in discussion on the Iowan erosion surface.

FIG. 3.7. Topography and geomorphology of area along South Turkey Creek, Adair County, Iowa. (A) Contours with interval of 10 feet. (B) Geomorphic surfaces with loess removed. Relate lettered locations to A.

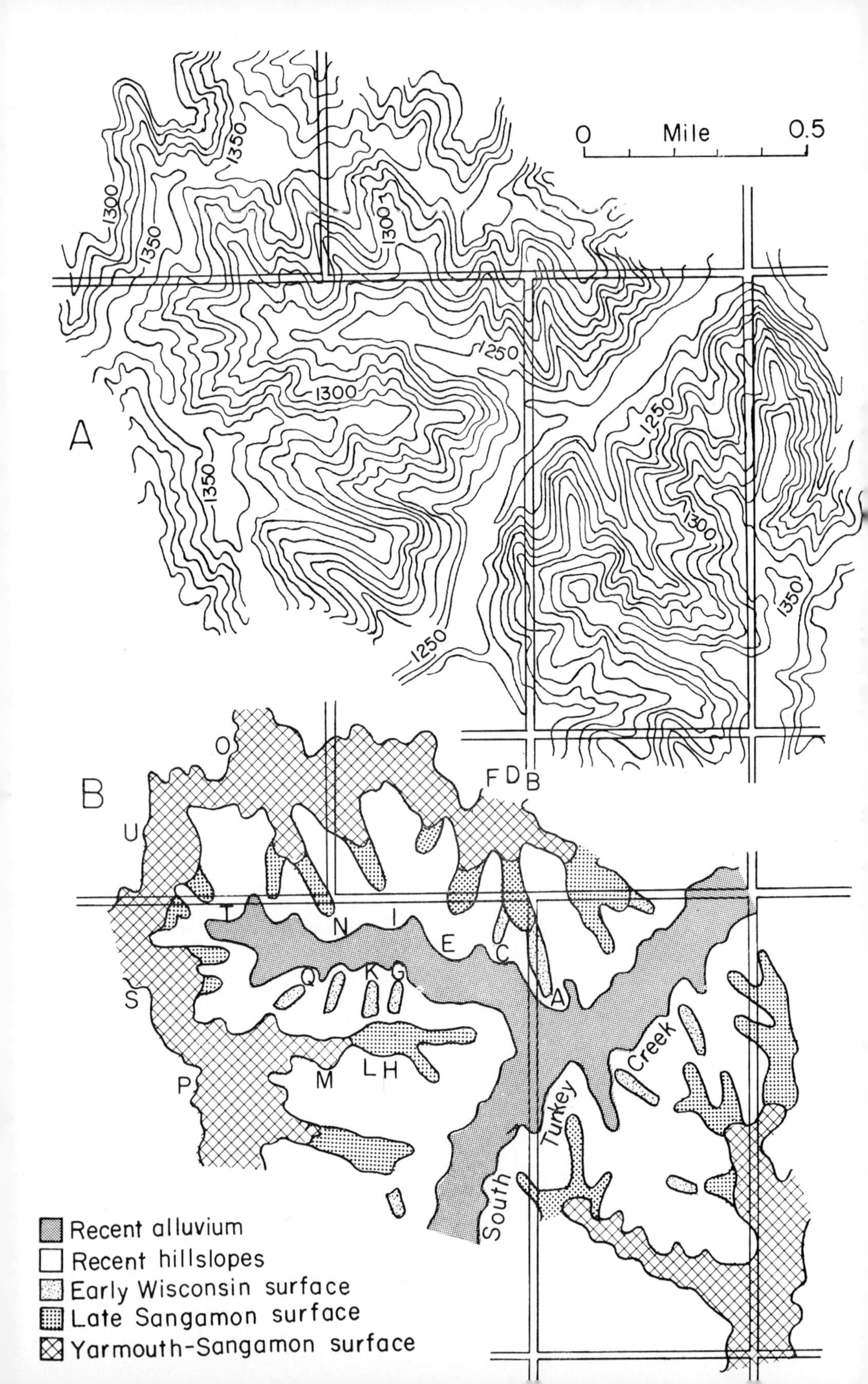

0
Mile
0.5
1350
1300
1350
1300
1250
1300
A
1350
1250
1300
1350
1250
B
O
F D B
U
T
N
I
E
C
Q
K
G
S
A
Creek
L H
M
P
Turkey
South
Recent alluvium
Recent hillslopes
Early Wisconsin surface
Late Sangamon surface
Yarmouth-Sangamon surface

What is the significance of the steps? The Yarmouth-Sangamon surface is a remnant of the Kansan drift plain that has remained and weathered since Kansan glaciation. The late Sangamon surface is an erosion surface that was cut into the Kansan till and below the Yarmouth-Sangamon surface and in the process removed the Yarmouth-Sangamon paleosol and also parts of the Kansan till. The Late Sangamon surface rises from a lower level along the valley slope to the higher Yarmouth-Sangamon surface (Fig. 3.8). The lowest level has the same kind of relationship to the Late Sangamon surface as the latter does to the Yarmouth-Sangamon surface.

What is the age of the lowest level? Wisconsin loess mantles the surface, but no paleosol intervenes. Therefore, deposition of loess must have rapidly followed the cutting of the lowest erosion surface so that time was not available to form a soil in Kansan till prior to burial by loess.

This lowest erosion surface, then, had to be cut during Wisconsin time and *during* the time of loess deposition. The surface is of Wisconsin age.

Local loess thicknesses are important in this analysis. The thicknesses on the Yarmouth-Sangamon and Late Sangamon surfaces are 15 to 16 feet. Maximum thickness on the Wisconsin surface is 7.5 feet. The hiatus represented by the stone line on the Wisconsin surface equates with 7.5 to 8.5 feet of loess. The erosion surface was cut during the earlier part of loess deposition in this area, and the 7.5 feet on the Wisconsin surface is the same uppermost 7.5 feet of the 15 to 16 feet on the adjacent Late Sangamon and Yarmouth-Sangamon surfaces.

A radiocarbon date from the basal soil of the Wisconsin loess at a depth of 15.5 to 16 feet and just above the Yarmouth-Sangamon paleosol is 18,700 ± 700 years (I-1411; see Table 2.3). The site is on the divide east of South Turkey Creek.

FIG. 3.8. Cross sections along interfluves *A* to *P* and *A* to *F* on map of Fig. 3.7. Geomorphic surfaces, paleosols, and deposits are identified.

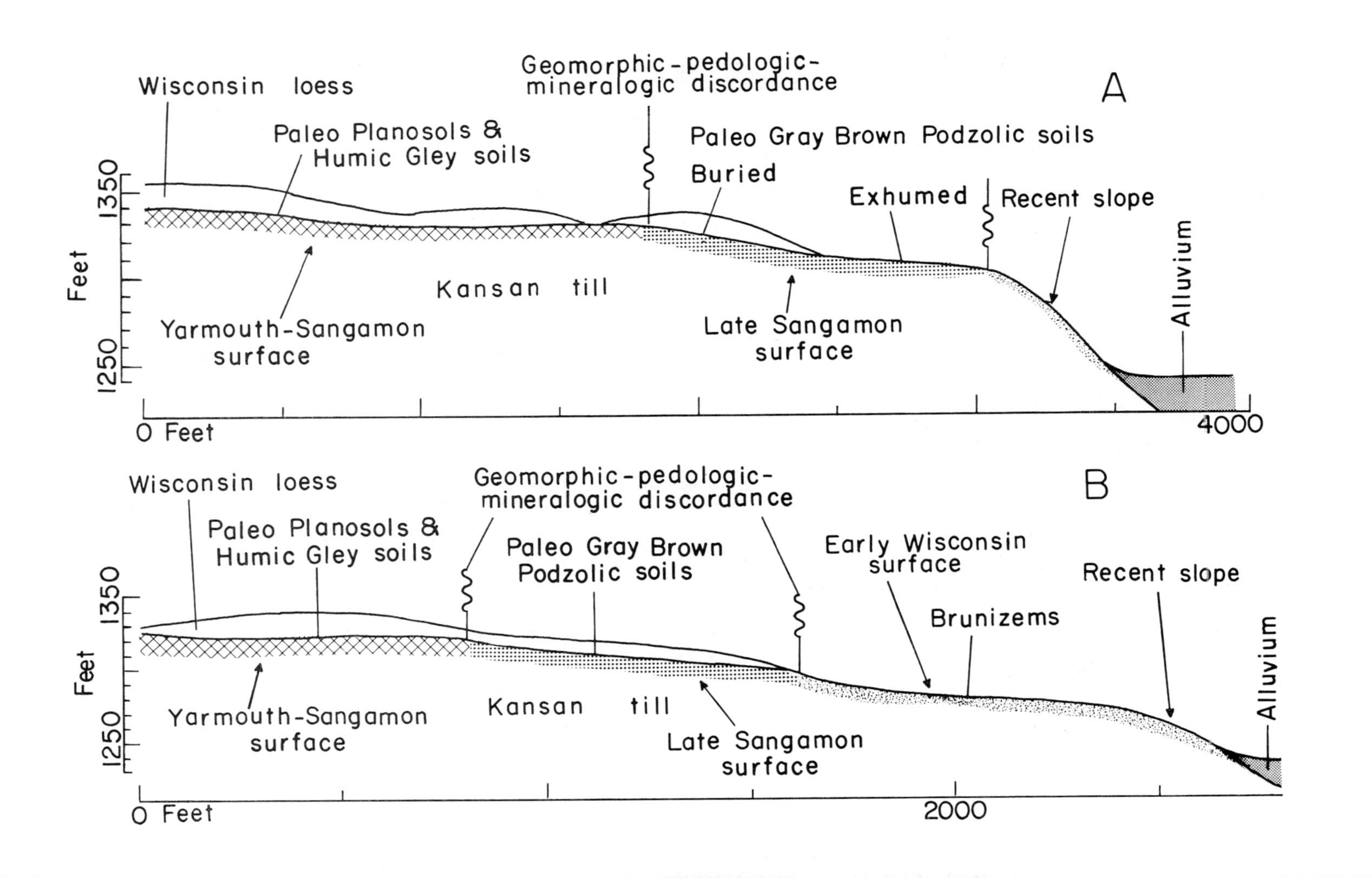
A
Wisconsin loess
Paleo Planosols &
Humic Gley soils
Geomorphic-pedologic-
mineralogic discordance
Paleo Gray Brown Podzolic soils
Buried
Exhumed
Recent slope
Alluvium
Kansan till
Yarmouth-Sangamon
surface
Late Sangamon
surface
Feet
1350
1250
0 Feet
4000
B
Wisconsin loess
Paleo Planosols &
Humic Gley soils
Geomorphic-pedologic-
mineralogic discordance
Paleo Gray Brown
Podzolic soils
Early Wisconsin
surface
Brunizems
Recent slope
Alluvium
Kansan till
Yarmouth-Sangamon
surface
Late Sangamon
surface
Feet
1350
1250
0 Feet
2000

We also know that the top of the loess is 14,000 years because that is the time of burial by Cary drift, as we recall. Thus, the Wisconsin erosion surface formed sometime between 14,000 and 18,700 years ago. This time in our general stratigraphic chronology in Iowa was during the Tazewell.

At some places the loess has been stripped from the early Wisconsin surface, so this exhumed surface was exposed to weathering since 14,000 years ago. Soils, then, could form in Kansan till and give rise to a third set of soils with a common parent material. Direct comparison of sets of soils within our chronology is made possible (Fig. 3.6). Following the same analysis that we applied to the Yarmouth-Sangamon paleosols, the Late Sangamon paleosols and the soils on the Wisconsin surface may be evaluated. Look at the differences. The Late Sangamon paleosols have yellowish-brown A2 horizons over reddish-brown B horizons in which clay has accumulated. These soils are analogous to Gray-Brown Podzolic soils that form under forest on the present surface. The Wisconsin surface soils have very dark gray A1 horizons over brown and yellowish-brown B horizons in which little if any clay has accumulated. These soils are Brunizems or soils that form under prairie.

The weathering indices (Fig. 3.6) show little significant mineral breakdown in the soils on the youngest surface, reasonable weathering in the Late Sangamon paleosols, and considerably more breakdown in the Yarmouth-Sangamon paleosols. In summarizing averages of solum thickness, B horizon thickness, clay content of B horizons, and weathering ratios, the differences between the soils on the three surfaces are very striking (Table 3.1). Each of these properties increases in magnitude from the youngest to the oldest surface and attests to the effect that time or duration of weathering has in the formation of soils. These data classically display a model for understanding the nature of soils in relation to geomorphic surfaces (Fig. 3.8).

During this discussion a Wisconsin erosion surface has been introduced which is characterized by a stone line on

TABLE 3.1. SET OF PROPERTIES OF THE SOIL-LANDSCAPE SYSTEM ON STEPPED EROSION SURFACES IN ADAIR COUNTY, IOWA

Soil	Thickness solum	Thickness B horizon	Clay content B horizon	Wrh	Wrl
	(in.)	*(in.)*	*(%)*		
Early Wisconsin surface					
M	29	22	34.6		
N	15	11	31.2		
O	32	23	32.2		
Averages	25	19	32.3		
A horizon				0.79	2.09
B horizon				0.92	2.12
C horizon				0.67	2.21
Late Sangamon surface					
H	39	29	49.5		
I	46	32	50.7		
J	70	56	49.1		
Averages	52	39	49.7		
A horizon				1.27	3.06
B horizon				1.04	2.49
C horizon				0.77	2.03
Yarmouth-Sangamon surface					
D	68	44	57.7		
F	85	62	50.7		
G	87	70	51.4		
Averages	80	59	53.2		
A horizon				2.05	4.64
B horizon				1.58	3.00
C horizon				1.21	2.48

Kansan till. Where buried under loess, no paleosol intervenes between the base of the loess and the till. This is an apropos point of departure to northeastern Iowa.

IOWAN EROSION SURFACE

An extensive area of 8,800 square miles in northeastern Iowa has previously been known as the Iowan drift region (Kay and Graham, 1941). The margin of this area has many lobes that protrude beyond a main body of the area (pl. 1). A considerable part of the area is covered by loess, but the major part is covered by a thin loam sediment that overlies a stone line on the till. The stone line, through the years, has been known as the Iowan pebble band.

The Iowan drift has a long history of controversy. In the earlier days, arguments involved the questions: (1) Is there an Iowan drift? (2) If it did exist, was it of pre-Wisconsin age, or was it a separate stage or a part of the Wisconsin? A related argument concerned the relation of loess and pebble band (stone line) to the drift. The questions involved: (1) Were the loess and stone line closely related in time to the drift, or (2) did the stone line between the loess and the till represent a hiatus of considerable time? McGee (1891), Calvin (1899), Alden and Leighton (1917), and Kay and Apfel (1929) developed the ideas through the years that the Iowan did exist and was a separate stage between the Illinoian and Wisconsin. Later, however, the Iowan was assigned to the earliest substage of the Wisconsin (Leighton, 1931, 1933; Kay and Graham, 1943). Accordingly, the stone line and loess were recognized as being closely related in time to the drift. On the other hand, Leverett (1909, 1926, 1939) contended: (1) that the Iowan was a late phase of Illinoian glaciation, (2) that the stone line had been formed by running water and much time was involved in its formation, and (3) that the loess was much younger than the drift. He (1942) later conceded that the Iowan was an early Wisconsin drift.

For the past two decades, existence of the Iowan drift has not been questioned, but an argument has arisen as to its placement in the Wisconsin stage. After formalizing the Farmdale, Leighton and Willman (1950) placed the Iowan as the next younger Wisconsin substage. However, all radiocarbon dates of wood extracted from the drift, which was identified by previous workers as Iowan, were "greater than" values, whereas the Farmdale is dated at 22,000 to 28,000 years. Accordingly, Ruhe and associates (1957, 1959) proposed and reaffirmed that the Iowan was older than Farmdale, but Leighton (1958, 1960, 1966) objected to this alignment.

We can examine the landscape problem in this part of the

state by starting from known areas and then crossing into the Iowan area. One study will cross from the known Kansan drift region to the Iowan and a second will cross from a paha onto the Iowan. A *paha* is a loess-capped prominence, elongated as ridges miles in length, or shortened to elliptical hills that stand apart on the Iowan plain or that merge with similar features to form lengthy ridges or broad plateaus (McGee, 1891; Fig. 3.9). The stratigraphy of the interiors of the paha is known to be very similar to that of the Kansan drift region (Scholtes, 1955).

So, we can cross the Iowan-Kansan border with a series of drill cores, with the first drilling site on the known Kansan area and the last on the Iowan area (Fig. 3.10). At site 1 the Yarmouth-Sangamon paleosol is overlain by 44 feet of loess, and the basal Wisconsin soil is present. The Yarmouth-Sangamon paleosol overlies the normal weathering zones of the Kansan till with a zone leached of carbonates above a zone that is calcareous. At site 2 the same general stratigraphic section is repeated. At sites 3, 4, and 5, the Yarmouth-Sangamon paleosol is progressively truncated. Wisconsin loess immediately overlies successively lower horizons of the paleosol. At site 6 the paleosol is missing, and from sites 7 to 14 the loess overlies either leached or calcareous Kansan till. These weathering zones in till can be traced in the cores right back across the border where they pass *under* the Yarmouth-Sangamon paleosol and therefore must be Kansan.

Where is the Iowan drift that has been written about all of these years? It does not exist. The Iowan plain at this place is an erosion surface cut into the Kansan till and below the level of the Yarmouth-Sangamon surface. From site 6 to 14, a stone line commonly separates the loess from the till. No paleosol separates the loess and till. Recall the relations of the Wisconsin surface relative to the Late Sangamon and Yarmouth-Sangamon surfaces in southern Iowa. The same principles apply here to the Iowan surface.

When did the Iowa erosion surface form? At site 1 the basal soil in Wisconsin loess is radiocarbon dated at 29,000

FIG. 3.9. Paha on the Iowan surface. (A) Panorama from Casey's paha, looking to northeast at lower-lying Iowan plain. (B) Core of the paha in road cut along defunct State Highway 402. Wisconsin loess (1) overlies Yarmouth-Sangamon paleosol (2) on Kansan till (3). Casey's paha is in the northeast corner of Tama County.

1
2
3

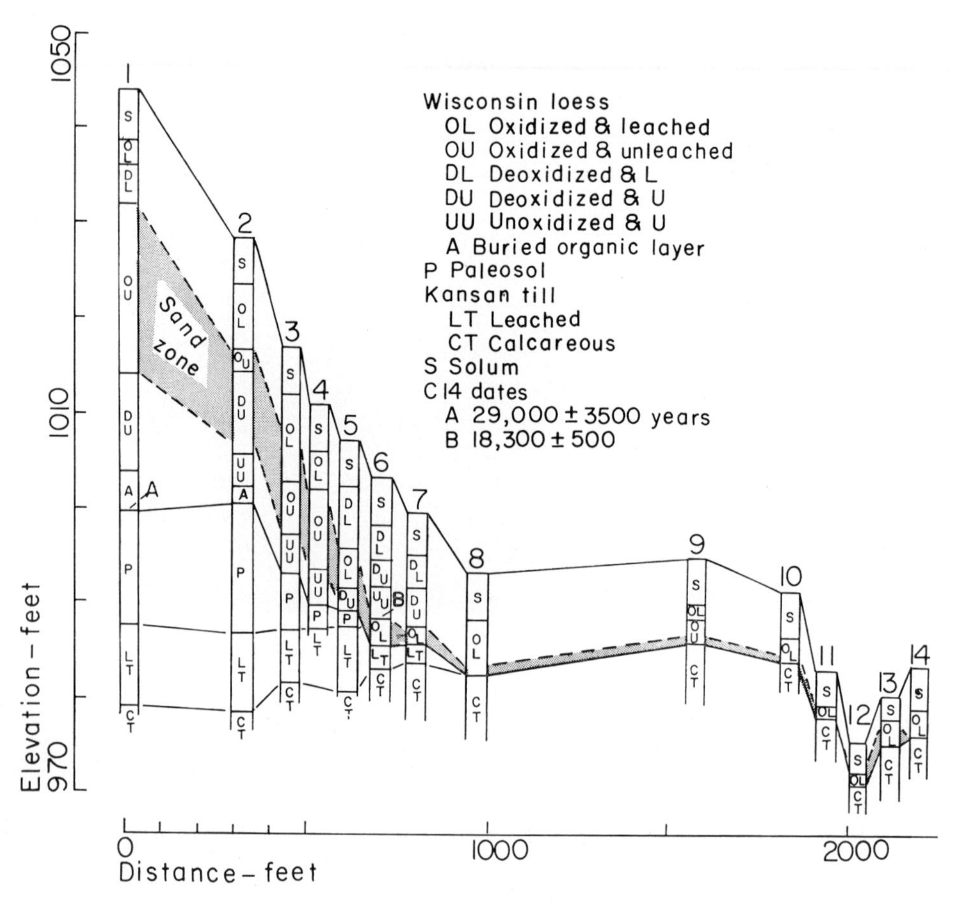

FIG. 3.10. Stratigraphy of drill cores across the Iowan-Kansan border at Salt Creek, Tama County. (Cf. plate 1 and Radiocarbon Catalog, Chapter 6.)

± 3,500 years (I-1269). At site 6 organic matter from the base of the loess that immediately overlies the erosion surface is 18,300 ± 500 years (W-1687). This dated sample is 700 feet to the north and 12 feet below the sample at site 1. Thus, the erosion surface formed between 18,300 and 29,000 years ago. This time span falls within the time of loess deposition in Iowa (Tables 2.3, 2.4).

The evidence is right here (Fig. 3.10). Note that a sand

zone immediately overlies the Kansan till from site 14 to site 6. Loess is above the sand zone but not below it. Note further that the sand zone flares upward into the loess from site 5 to site 1. Why? Sand blew from the exposed erosion surface onto the higher upland prior to loess burial on the lower surface. Sand could not move in the opposite direction because there is no available source on the highland. The loess below the sand zone at sites 1 to 5 is relatively sand-free. What does this mean? Loess was being deposited in the area and began 29,000 years ago. At some later date but prior to 18,300 years ago, the Iowan erosion surface formed in till and supplied sand to be added to the loess which continued to be deposited. The erosion surface stabilized, and the sand supply was shut off. Loess deposition continued and blanketed the area, burying the sand zone on the highland and the lower lying erosion surface. The last loess increment covers the whole area. The earliest increment, below the sand zone at sites 1 to 5, equates with the hiatus represented by the stone line of the erosion surface on Kansan till.

These relations are repeated over and over again at paha within the Iowan area. At 4-Mile Creek paha in Tama County (pl. 1; Radiocarbon Catalog, Chapter 6), the argument is further clinched by the dual presence of the Yarmouth-Sangamon and Aftonian paleosols that permits accurate reconstruction of the subsurface stratigraphy (Fig. 3.11). At Hayward's paha in northeast Tama County and following the identical principles used at Salt Creek, the Iowan erosion surface was cut between 25,000 ± 2,500 years (I-1267) and 20,000 ± 400 years ago (I-1409; see Table 3.2). At Alburnett paha in Linn County, the surface was cut between 20,700 ± 500 years (I-2332) and 12,700 ± 290 years ago (I-2333). The Alburnett study is particularly important because dating of the broad expanse of loam sediments on the Iowan surface (pl. 1) is determinable here. The loam-sediment landscape is younger than 12,700 years because it is the lower lying plain that extends beyond this younger sample locale. This means that this large area is younger than the Cary drift of the Des Moines lobe (Table

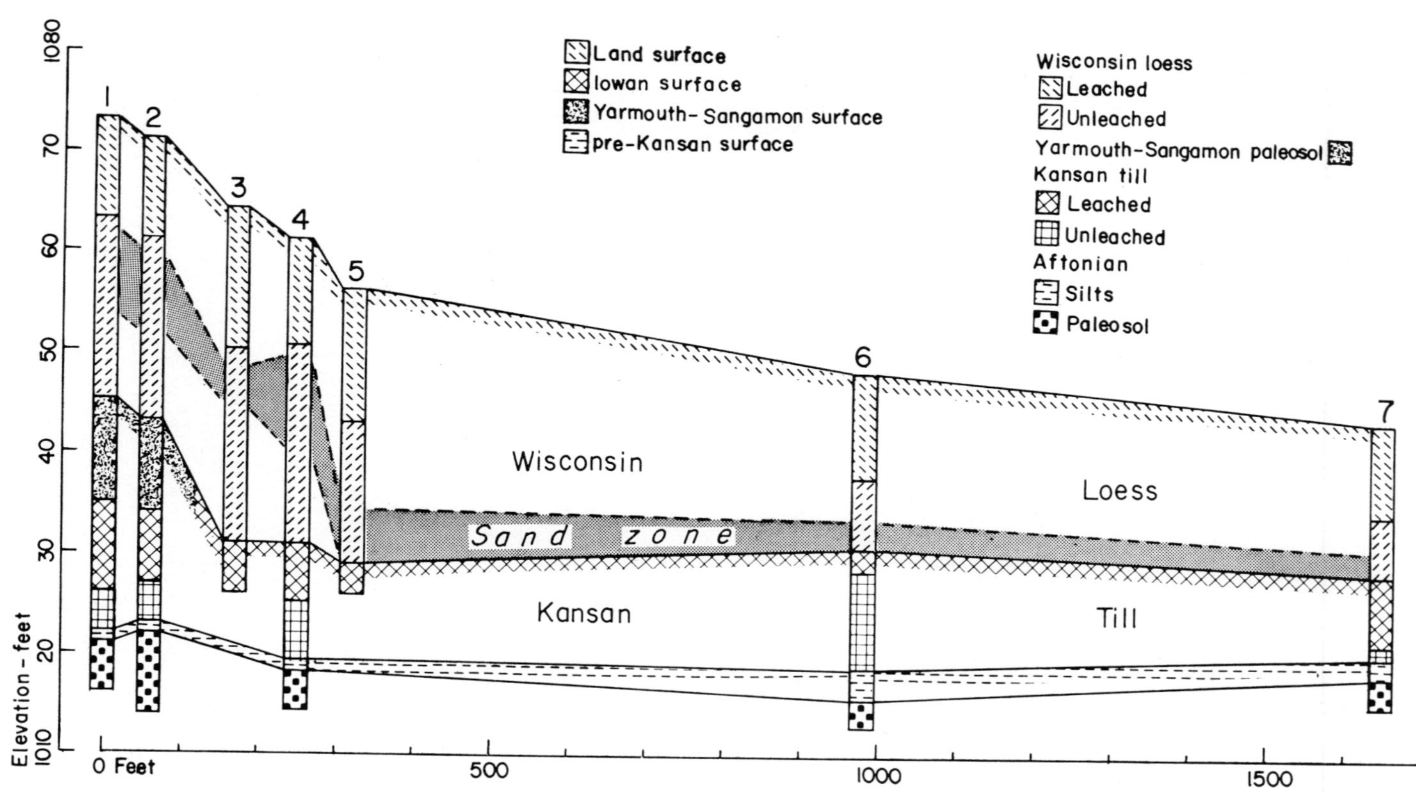

FIG. 3.11. Stratigraphy of drill cores from 4-Mile Creek paha onto lower lying Iowan plain.

TABLE 3.2. RADIOCARBON DATES ON THE IOWAN EROSION SURFACE IN IOWA

Sample*	Date in years before present	Location	Notes
I-1860	2,930 ± 110	Sumner bog, Bremer County	Peaty muck at 27- to 30-inch depth; grass-herb pollen zone
I-1861	6,130 ± 120	Sumner bog, Bremer County	Muck at 54- to 57-inch depth; tree-grass transition pollen zone
I-1862	11,880 ± 170	Sumner bog, Bremer County	Muck at 88 to 91 inches at base of bog; conifer pollen zone; larch wood
I-2333	12,700 ± 290	Alburnett paha, Linn County	Spruce wood from peat
W-1687	18,300 ± 500	Salt Creek, Tama County	Soil organic carbon (OC), residue
I-2329	18,400 ± 310	4-Mile Creek, Tama County	OC from alluvuim below Iowan plain
I-1409	20,300 ± 400	Hayward's paha, Tama County	OC, residue
I-2332	20,700 ± 500	Alburnett paha, Linn County	OC, residue
W-1681	21,600 ± 600	Palermo area, Grundy County	OC, residue
I-1404	22,600 ± 600	Palermo area, Grundy County	OC, residue
OWU-167	23,050 ± 820	Wapello, Louisa County	Peat below upper Lake Calvin terrace sediments
I-1805	23,750 ± 600	Wapello, Louisa County	Peat below upper Lake Calvin terrace sediments
I-1267	25,000 ± 2,500	Hayward's paha, Tama County	OC, residue
I-1269	29,000 ± 3,500	Salt Creek, Tama County	OC, residue
I-2330	34,900 + 2,700 − 1,700	Elma, Howard County	OC, residue
W-880	37,600 ± 1,500	Logan, Harrison County	Spruce wood from terrace alluvium

* Sample numbers are I for Isotopes, Inc.; OWU for Ohio Wesleyan University; and W for U.S. Geological Survey, Washington, D.C.

2.7). In fact, some of the Iowan surface is much younger and of quite recent age. Hillslopes around Sumner bog in Bremer County descend to a bog fill that is 11,800 years old at the base, 6,130 years old near the middle of the fill, and 2,930 years old a few feet below the surface of the bog. The hillslopes can be as young as the two younger radiocarbon dates.

What are paha? They are not topographic highs of the Kansan drift around which "Iowan glacier ice" flowed but did not cover, as previously believed (Scholtes, 1955). Instead, the cores of the paha are nothing more than erosion remnants of older land surfaces standing above the Iowan erosion surface at interstream divides. The erosion surface encroached on these areas from all directions but did not completely obliterate the old surfaces. As remnants, they received the full deposition of loess in contrast to the lower lying erosion surface which did not.

"Greater Than" Dates

In radiocarbon dating, values that are beyond reach of the method are reported as "greater than (>)." This means that a statistical counting of disintegrations per minute does not fall within a standard deviation of the count. In other words, the half-life of the half-life and so on of the exponential decay rate is well along the path toward an asymptotic oblivion. These samples are old or, in the jargon, "dead." A number of such dates became embroiled in the argument on the Iowan "drift." Because all samples from so-called Iowan drift always came out dead (Table 3.3), the Iowan was considered old and certainly older than Farmdale which is datable at 22,000 to 28,000 years (Ruhe *et al.*, 1957, 1959). However, the argument still goes on that the "Iowan glacier ice" continually dug up old wood rather than bulldozed trees in its advance (Leighton, 1958, 1960, 1966).

Well, the first inconsistency in this latter argument is that Iowan drift does not exist. A more damaging refutation is at

TABLE 3.3. OLD RADIOCARBON DATES FROM KANSAN AND NEBRASKAN TILLS BENEATH IOWAN SURFACE*

Sample†	Date in years before present	Location	Notes
W-503	> 29,000	Fayette, Fayette County	Hemlock wood
I-1265	> 30,000	Palermo area, Grundy County	Soil organic carbon (OC) between upper and lower tills
W-534	> 34,000	Fayette, Fayette County	Hemlock wood
W-514	> 35,000	Scranton No. 1, Greene County	Spruce wood in lower till
W-516	> 35,000	Maynard, Fayette County	Spruce wood
I-1405	> 36,000	Palermo area, Grundy County	OC between upper and lower tills
W-591	> 37,000	Quimby, Cherokee County	Larch wood
W-599	> 37,000	Central City, Linn County	Spruce wood
W-600	> 37,000	Independence, Buchanan County	Hemlock wood in peat between tills
W-139	> 38,000	Independence, Buchanan County	Hemlock wood in silt between tills
I-1863	> 39,900	Mill Creek, Cherokee County	Spruce wood in lower till
I-2758	> 39,900	Fairview,‡ Sioux County	Larch wood
I-1025	> 40,000	Madrid,§ Polk County	Spruce wood in lower till
I-1266	> 40,000	Palermo area, Grundy County	OC between upper and lower tills

* See plate 1 and Radiocarbon Catalog, Chapter 6, for locations and descriptions.

† Sample numbers are I for Isotopes, Inc.; and W for U.S. Geological Survey, Washington, D.C.

‡ Fairview is just across the state line (Big Sioux River) in South Dakota.

§ Madrid is in Boone County but site is just across county line in Polk County.

Alburnett paha. The two tills in that paha core are Kansan and Nebraskan. The lower Iowan plain north of the paha is cut in the Nebraskan till. This till can be traced directly across country a few miles to the Central City road cut in Linn County. There spruce wood from this till is > 37,000 years old (W-599; see Table 3.3), and no wonder; it is Nebraskan in age. Obviously, it is not Iowan "drift." At all other "greater than" localities, the story is similar.

Iowan Erosion Surface of Northwest Iowa

The fact that Iowan drift does not exist in northeastern Iowa requires reevaluation of the Iowan drift region of northwestern Iowa (Smith and Riecken, 1947; Ruhe, 1950, 1952). Relations there (pl. 1) are much the same as in the other corner of the state. We must return to the previous assessment that the region is Kansan and that the "Kansan gumbotil" was stripped by erosion (Kay, 1917; Carman, 1917, 1931; Kay and Apfel, 1929).

Two radiocarbon sections bear on this problem. Just east of the Big Sioux River and Fairview, South Dakota, but in Sioux County, Iowa, larch wood from the base of the upper till is > 39,900 years (I-2758). Just east of Mill Creek in Cherokee County, spruce wood from the lower till is > 39,900 years (I-1863). Both of these tills should be Iowan drift according to the studies of 1947 to 1952. The radiocarbon ages conform to the suite of ages of other samples beneath the Iowan surface (Table 3.3). The tills are not Iowan but probably Kansan.

Lake Calvin of Southeast Iowa

Recognition of the Iowan erosion surface in northeastern Iowa also raises problems concerning Lake Calvin in southeastern Iowa (pl. 1). Lake Calvin previously has been related to Illinoian glaciation (Schoewe, 1920). The conformance of the eastern margin of the lake beds to the Illinoian drift border led to a belief that Illinoian glacier ice had dammed the lower reaches of ancestral Iowa and Cedar rivers and caused waters to back up the valleys, forming the lake. However, problems in terrace correlation between the lower Iowa River Valley and Lake Calvin led Shaffer (1954) to suggest that the high terrace of Lake Calvin was not Illinoian but Tazewell.

Radiocarbon dates confirm Shaffer's study. A peat that is buried by the high terrace alluvium near Wapello in Louisa County is 23,050 ± 820 years (OWU-167) and 23,750 ± 600 years old (I-1865; see Table 3.2). The high terrace alluvium is above this peat and has to be younger. The dates are Farmdale in the general Pleistocene chronology and younger ages fall within the Tazewell. But more importantly, these dates fall within the time span of the Iowan erosion surface upstream in both the Iowa and Cedar rivers watersheds. Eroded debris had to be transported down river. Consequently, the Lake Calvin terraces are down-valley depositional components of the Iowan erosion surface and are not related to Illinoian glaciation which is much older.

Alluvial Fills

Two other alluvial fills in valleys also relate to Iowan erosion surface formation. In Tama County along 4-Mile Creek, which heads on the Iowan surface, organic carbon in alluvium at a depth of 14 to 15 feet is 18,400 ± 310 years old (I-2329; see Table 3.2). The Iowan surface was cut in the near vicinity at Salt Creek at 18,300 ± 500 years ago (W-1687).

Along the Boyer River near Logan, Harrison County, spruce wood from alluvium buried under Wisconsin loess is 37,600 ± 1,500 years old (W-880; see Table 3.2). The base of the Wisconsin loess is dated by spruce wood at 19,050 ± 300 years (W-879; see Table 2.3). The site is a loess-mantled terrace along the Boyer River which currently heads on the Tazewell drift of northwestern Iowa (Corliss and Ruhe, 1955; Daniels and Jordan, 1966). This stream also should have headed on the Iowan erosion surface prior to Tazewell glaciation. The alluvium probably is a down-valley depositional phenomenon related to early Iowan erosion surface formation.

TAZEWELL DRIFT OF NORTHWEST IOWA

One last item beneath the Wisconsin loess requires coverage and the picture in the state will be completed. The Tazewell drift was delineated in parts of Lyon, Osceola, Dickinson, O'Brien, Clay, Cherokee, Buena Vista, Ida, and Sac counties (Ruhe, 1950, 1952). This drift region is essentially the same one that earlier had been called the "Iowan" (Carman, 1931).

A question arises now with the decease of the Iowan. Does the Tazewell exist? Yes. Not only does an identifiable till sheet bury loess, but it is, in turn, buried by loess which also is capped by the Cary till. Further, radiocarbon dates substantiate the separation of the Tazewell. At two localities, one near Sheldon in O'Brien County and the other near Cherokee in Cherokee County (pl. 1), radiocarbon dates from the Tazewell till are 20,000 years. At the Sheldon site the section from the surface downward is loess, 27 inches; calcareous till, 32 inches; calcareous loess, 14 inches; buried soil A horizon, 3 inches; buried soil B horizon to depth, 15+ inches. Organic carbon from the buried A horizon is 20,500 ± 400 years old (I-1864A). The till above this buried A horizon is younger.

In a road cut along State Highways 3 and 5 on the east valley slope of Little Sioux River near Cherokee, spruce wood from near the base of the exposed till is 20,000 ± 800 years old (O-1325). Note the agreement in age between these two dates. Both sites are at the margin of the mapped Tazewell drift (pl. 1). The dates also fall into Tazewell time at the type area in Illinois (Frye and Willman, 1960). So, there is a Tazewell drift in Iowa.

Surface Modification

But now recall that in our developing radiocarbon history, two important events were evolving about this time. Loess deposition was going on from 29,000 to 14,000 years ago, and the Iowan erosion surface also was forming (Table 3.2). Have these things affected the Tazewell drift? They certainly have. Not only does the Tazewell till bury loess such as at the Sheldon section, but the drift surface has a continuous loess mantle that buries it. This would be expected as loess deposition was only interrupted by Tazewell glaciation in the northwestern part of the state.

The Tazewell surface has also been modified by extension of the Iowan erosion surface on it. This may be illustrated by examining the drainage nets in a county-wide transect from Cherokee to Pocahontas County. But before doing so, some jargon about streams must be introduced. A *drainage net* is the geographic pattern that a stream system forms on the landscape. There are various kinds. The most common net is *dendritic* in which tributaries join a main stream with acute angles and the pattern is treelike. If main and tributary streams join at right angles, the net is *rectangular*. Where a rectangular pattern has several systems of nearly parallel trunk streams, the net is *trellis*. Tributaries join the main stream at right angles, but other junctions are also at right angles and the stream courses parallel the main stream. In *barbed* nets, tributaries join main streams in hooked bends that point upstream. *Radial* nets originate at a common center and diverge outward, but conversely, *centripetal* nets originate at a peripheral area and converge toward a common center.

In order to handle various parts of the drainage net, one resorts to stream ordering, and the order is determined by the branching or bifurcation of a stream. In the Horton (1945) system, subsequently modified by others, the smallest fingertip side streams in a net are order 1; order 2 streams receive order 1; order 3 receive order 2, and so on. The main stream is assigned an order based on counting down from the farthest unbranched tributary.

A natural outgrowth of the branching or bifurcation of the main system is the density of the drainage net. This may be defined as the total length of streams within a drainage basin per unit of area ($Dd = \Sigma L/A$). Measuring the total lengths of streams in a well-developed drainage basin is a formidable task. The job may be simplified by using a point intercept method and employing an electric grid counter. An area grid with copper wires imbedded in plastic is laid over a map of a drainage net which is then traced on the plastic. The tracing pencil is tied into the electric system. At each intersection of a drainage line and a copper grid line, an electric contact is made or a point intercept is recorded cumulatively by an electric counter. The area of the drainage net is also rapidly determined by the fit of squares of the electric grid. Drainage density then is $Dd = \Sigma P/A$ where P is total point intercepts.

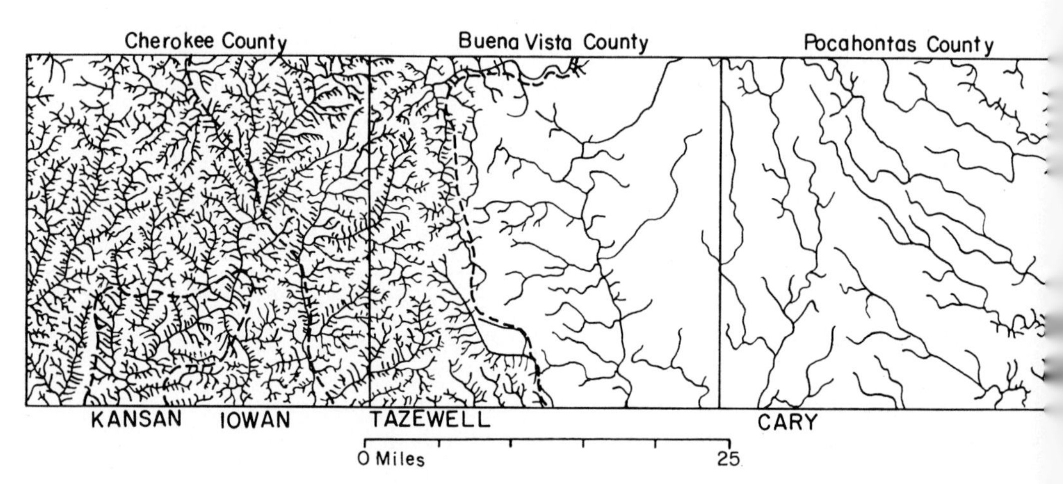

FIG. 3.12. Drainage nets on Kansan area, Iowan erosion surface area, and Tazewell and Cary drift areas in northwest Iowa.

With this background we can return to the problem of the Tazewell surface in northwest Iowa. In the county wide transect the nets are dendritic, and there are obvious differences in drainage density from the west to the east (Fig. 3.12). To quantify these differences, two samples, each of 144 square miles, are drawn for each of the Kansan, Iowan, Tazewell, and Cary areas. Drainage densities by point intercept are respectively: 10.2, 7.9; 7.7, 6.1; 5.4, 4.7; and 2.1, 1.9. The densities progressively decrease from the Kansan area to the Cary area.

Samples of stream orders have been established for these areas (Leopold, Wolman, and Miller, 1964). Using their data of stream orders, comparisons may be made with drainage density by point intercept and with time based on radiocarbon dating (Fig. 3.13). As expected, drainage density increases as the number of first-, second-, and third-order streams increases, and the relation is exponential. As the drainage net extends headward and integrates within the system, the first-order streams become more abundant than second order which, in turn, are more abundant than third order, and so on. A basic geomorphic principle is brought out here. The development of an area into a watershed or drainage basin is done mainly by first- and second-order streams, that is, side valley and side streams.

The drainage net development can be placed within a time framework by inserting radiocarbon dates at pertinent places within the system. First, note that a distinct break occurs in the drainage density at the margin of the Cary drift (Fig. 3.12). On the Cary surface the density values are 1.9 and 2.1. On the immediately adjacent Tazewell surface the values are 4.7 and 5.4. The density increases from Cary to Tazewell by factors of 2.5 and 2.6. As expected, there are greater numbers of first-, second-, and third-order streams on the Tazewell than on the Cary surface (Fig. 3.13). This means that drainage net extension and integration was well advanced on the Tazewell surface at the time of Cary glaciation. Since we know that this glaciation began 14,000 years ago and was essentially completed in Iowa 13,000 years ago (Table 2.7), most of the drainage net must have been established on the Tazewell sur-

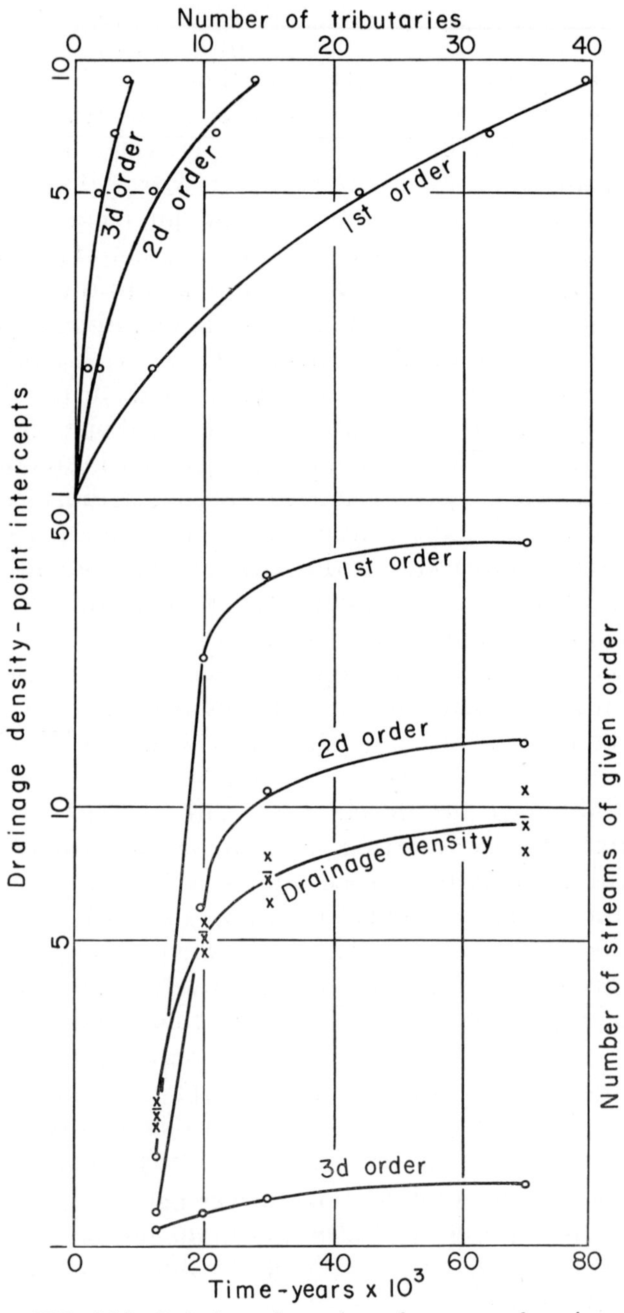

FIG. 3.13. Relation of number of streams of a given order to drainage density and of drainage density and number of streams of given order to time. All curves, Cary at left to Kansan at right.

face prior to those times. These data also show that relatively little of the drainage net developed since then.

Now, compare the drainage nets on the Tazewell surface and the adjacent Iowan erosion surface (Fig. 3.12). They appear to be very similar, but quantitatively they differ. On the Tazewell surface the values are 4.7 and 5.4, but on the Iowan surface the values are 6.1 and 7.7. The density changes across the border by factors of 1.3 and 1.4. There are greater numbers of first-, second-, and third-order streams on the Iowan than on the Tazewell surfaces, but the differences are much less than those between Tazewell and Cary.

The slighter differences between Iowan and Tazewell mean that drainage net development was somewhat advanced on the Iowan surface at the time of Tazewell glaciation about 20,000 years ago (I-1864A and O-1325). But, soon after uncovering of the Tazewell drift by glacier ice, the drainage net extended and integrated on the Tazewell surface, resulting in the similarity of density patterns on the two surfaces. By the radiocarbon dates, the net on the Tazewell surface formed mainly between 20,000 and 14,000 years ago. The pattern on the Iowan formed prior to 20,000 years ago (Table 3.2). However, the patterns together show that the drainage net on the Iowan erosion surface extended headward across the Tazewell drift and integrated on that drift surface after 20,000 years ago. This means that the Iowan erosion surface extends across the Tazewell drift in northwest Iowa. This should be expected, as dates on the Iowan surface in northeast Iowa are as young as 18,300 years and younger (Table 3.2).

A last point on this matter involves uncontrolled use of morphometric analysis. Just resorting to an analysis of drainage density and number of streams of given order versus time (Fig. 3.13), a conclusion may be reached that "the rate of drainage development is most rapid in the first 20,000 years and subsequently levels off to slower rate" (Leopold, Wolman, and Miller, 1964, p. 424). The curves per se are misleading as they show relatively rapid development of drainage to 20,000 years ago and a leveling off to a slower rate before that time. But, field knowledge of the system shows that drainage development was relatively rapid 14,000 to 20,000 years ago but was relatively slow since 14,000 years ago. This contradiction illustrates the danger of relying on morphometric analysis without field knowledge.

Glacial Episodes of the Wisconsin in Iowa

We have now reviewed two major glacial episodes of the Wisconsin in Iowa, the Tazewell and the Cary, and they are all that are present. This is a major revision in the Pleistocene part of the Quaternary of Iowa, and this revision must be placed in historical context.

The work of Kay and his associates resulted in the recognition of the Iowan glacial, Peorian intraglacial, and Mankato glacial episodes. The Peorian was the episode of loess deposition. In the late forties and early fifties, Iowan, Tazewell, Cary, and Mankato glacial episodes were recognized with loess associated with the Iowan and Tazewell. Around 1960 the Mankato was dropped as no longer recognizable, and in 1965 the Iowan was dropped as nonexistent. We are left with the Tazewell and Cary.

At present we recognize the Wisconsin deposits as an informally designated Wisconsin loess that is dated at 30,000 to 14,000 years ago, the Tazewell drift that is dated at 22,000 to 14,000 years ago, and the Cary drift that is dated at 14,000 to 13,000 years ago. Less than 13,000 years ago is Recent.

All of the foregoing features are the foundation on which the present landscape of Iowa is built. This leads us to the land surface.

CHAPTER 4

THE LAND SURFACE

IT WILL NOT BE POSSIBLE TO EXAMINE all of the land surface of Iowa without ballooning the length of our discussion beyond reason. Consequently, we will handle the problem by looking at landscape models that are characteristic of large parts of the state. We can do this by studying hill summits, hillslopes, and valley features within the broad regions of the loess and glacial drift terrains (pl. 1). Soils are an essential part of any landscape, so they will be introduced as needed. This will lead to formulation of landscape models and soil systems. The principles developed can be applied readily, not only in other parts of the state but in similar provinces in other parts of the United States and abroad as well.

First, let us look at the tops of the hills and recall that the "when" of the feature is our major interest.

HILL SUMMITS

With the qualification that excludes the outcrops of paleosols, the oldest hilltops or summits on the present land surface of Iowa are 14,000 years old. These summits were constructed by deposition of eolian silt in the loess province of the state. The next oldest hilltops were constructed by deposition of glacial till of the Cary drift of the Des Moines lobe. Some of these summits may be as old as 13,000 years.

Loess Province

At this point we should recall but not repeat the discussion concerning the properties of loess in relation to the Missouri River Valley source (Chapter 2). This loess province will be our model for the formulation of a regional picture. Recall that the loess thins from the source toward the southeast, that the content of coarser particles decreases, that the content of finer particles increases, and that the absolute age of the base of loess is progressively younger. All of these relations are so systematic that they are expressible mathematically. Recall further that the loess was deposited on a preexisting land surface that had hills and valleys. Consequently, a lid was placed on the landscape by deposition of the Wisconsin loess, and the summits of the present hills were built up by the eolian silt.

Is there some sort of system in these summits? Starting near the Missouri River Valley, the summits are relatively narrow, being a tenth of a mile or less in width, and stand relatively high above the adjacent valleys. This local relief ranges from 150 to 200 feet. To the southeast, the widths of summits progressively increase, and the relief from hilltop to valley progressively decreases. These relations are so systematic that they may also be expressed mathematically (Fig. 4.1.). Summit width (Y) increases linearly as distance (X) increases across the region to south central Iowa as expressed by:

$$Y = 0.019 + 0.0076X$$

Relief between summit and valley (Y) curvilinearly decreases as distance (X) increases as:

$$Y = 1/(0.0034 + 0.00012X)$$

We can summarize at this point. As the loess thins (1) and particle size decreases (2), the summits broaden (3) and the local relief decreases (4). Further, the paleosols on which the loess was deposited are at progressively shallower depths, or the more impermeable paleosols (Chapter 3) become closer to the land surface beneath the summits (5).

Recall again that two deoxidized zones in the loess merged eastward away from the Missouri River Valley (Chapter 2). These zones indicate relict gleying or zones of water saturation that were perched on the buried paleosols. The depth to the deoxidized zone also systematically decreases as the loess thins, particle size decreases, summits broaden, and local relief decreases. The depth to the relict saturation zone (Y) decreases with distance (X) as expressed by:

$$Y = 434.9 - 173.7 \log X$$

The slope of this curve is more gradual than that of the loess thickness curve (Fig. 4.2). A direct relation must then exist

FIG. 4.1. Widths of loess summits increase linearly with distance from the Missouri River Valley in western Iowa to south central Iowa. Maximum local relief from summits to adjacent valleys decreases curvilinearly with distance.

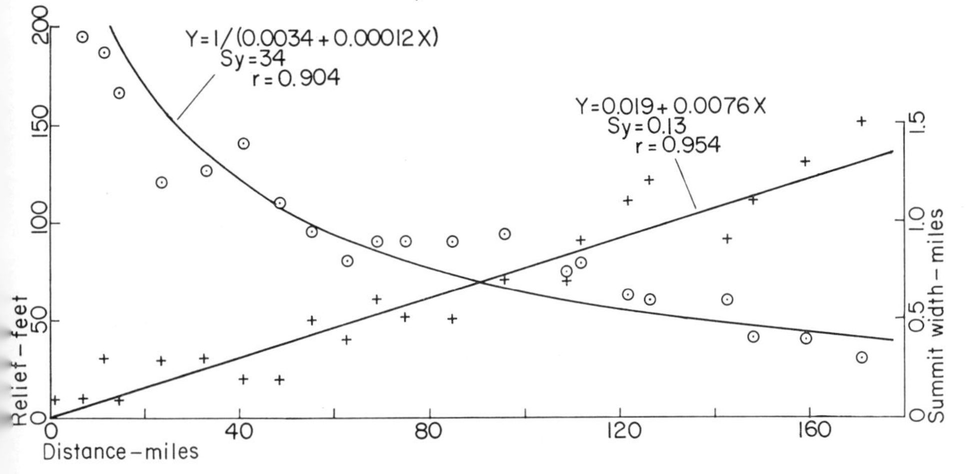

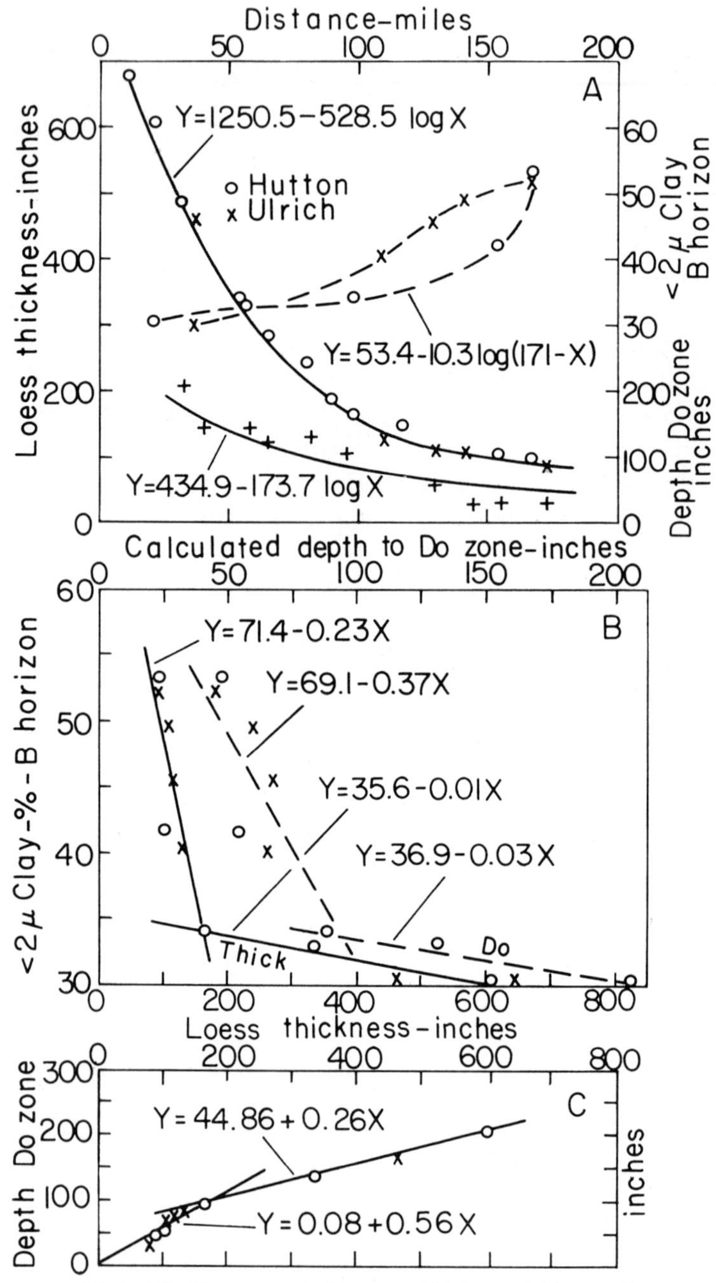

FIG. 4.2. From top to bottom, (A) loess thickness decreases with distance from Missouri River Valley in western Iowa to south central Iowa. Clay content of B horizons of soils increases with distance. (From Hutton, 1947, and Ulrich, 1950.) Depth to deoxidized zone in loess decreases with distance. (B) Bipartite distributions, showing increase of clay content of B horizons of soils as depth to deoxidized zone decreases and loess thickness decreases. (C) Bipartite distribution, showing decrease to depth of deoxidized zone as loess thickness decreases.

between the loess thickness *(X)* and the depth to the deoxidized zone *(Y)*, and it is a bipartite relation. Where loess thickness is more than 160 inches, one part of the relation is expressed as:

$$Y = 44.86 + 0.26X$$

Where loess thickness is less than 160 inches, the other part of the relation is expressed by:

$$Y = 0.08 + 0.56X$$

The latter curve almost passes through the point of origin, checking the exactness of fit (Fig. 4.2). Expressed in another way, these relations mean that with distance the deoxidized zone or perched zone of saturation is progressively closer to the land surface and any solum of a soil formed on the surface.

A pattern of soil properties is related to the regional pattern of loess thinning. In southwest Iowa in soils formed under grass in conditions of better internal drainage (Hutton, 1947, 1951) and in soils in conditions of poorer internal drainage (Ulrich, 1950, 1951), soil development increases as the loess thins. Among many other properties, the clay content in the soil B horizons progressively increases as the loess thins (Table 4.1). This property related to loess thickness also is a bipartite linear distribution (Fig. 4.2). Where loess thickness is more than 160 inches, clay content *(Y)* versus thickness *(X)* is expressed as:

$$Y = 35.63 - 0.0096X$$

Where loess thickness is less than 160 inches, clay content is expressed by:

$$Y = 71.37 - 0.228X$$

These equations are the sixth functional relation that we have formulated in our system.

TABLE 4.1. INCREASE IN CLAY CONTENT OF B HORIZONS OF SOILS RELATED TO LOESS THICKNESS IN SOUTHWEST IOWA

Loess thickness	$< 2\mu$ Clay—B horizon	Soil	Kind	Location
(in.)	*(%)*			*(county)*
603	30.4	Monona silt loam	Prairie	Monona
468	30.2	Minden silt loam	Humic Gley	Shelby
338	33.0	Marshall silt loam	Prairie	Shelby
169	34.3	Sharpsburg silty clay loam	Prairie	Adair
130	40.1	Winterset silty clay loam	Humic Gley	Union
112	45.5	Haig silty clay loam	Humic Gley	Clarke
105	49.5	Haig silty clay loam	Humic Gley	Decatur
102	41.6	Grundy silty clay loam	Prairie	Decatur
95	53.2	Seymour silty clay loam	Prairie	Wayne
90	52.4	Edina silt loam	Planosol	Wayne

Source: Hutton, 1947; and Ulrich, 1950.

Now, if depth to upper or merged deoxidized zone is related to loess thickness, and amount of clay in soil B horizons is related to loess thickness, then there must be a relation between B horizon clay content and depth to the deoxidized zone. There is, but again the relation is a set comprising two linear subsets (Fig. 4.2). The discordance between these subsets is at the depth of 90 inches to deoxidized zone. Where the depth of this zone is less, the clay content of B horizon (Y) relates to depth of deoxidized zone (X) as:

$$Y = 69.12 - 0.375X$$

Where the depth is greater, the relation is:

$$Y = 36.95 - 0.034X$$

This means that as the solum of the soil becomes progressively closer to the relict saturation zone, the amount of B horizon clay becomes progressively much greater. Consequently, the soils should have been in a wetter environment that could lead to greater weathering and formation of clay in the soil.

Topographically, the heavier textured soils coincide with the region of broad, tabular divides that are shallowly underlain by extremely clayey paleosols, the relict "gumbotil plain" of the Pleistocene. This is also the region where the local relief is lowest (Fig. 4.1). Thus, internal soil drainage is impeded in downward percolation by the paleosols and in lateral movement by the broad summits and low relief. These stratigraphic and topographic conditions give rise to a more moist soil environment that has occurred and does occur in this region. To the westward, the saturation zone was and is much deeper as the paleosols are deeper, the ridges are narrower, and relief is greater. Internal soil drainage is not impeded, and soil weathering is not so intense.

We can validate the conclusion of soil environment another way. Using volume weight (weight per unit of volume of an undisturbed soil sample), porosity (percentage of volume occupied by voids), and soil solids (percentage of volume occupied by soil solids), the height of a soil column filled by rainfall percolation may be estimated (Table 4.2). The data are for the C horizons or parent materials of the soils. As the model explains soil formation, the solum must be excluded. In other words, the system at the beginning of weathering must be evaluated.

In addition to this premise, calculations are for optimum conditions under present rainfall that: (1) All soil voids are filled with water. (2) All water perches in the loess above the more impermeable paleosols. (3) No water is lost through surface runoff. (4) No water is lost through evapo-transpiration. For the Minden soil a zone of saturation 58 inches thick could perch above the paleosol, but the water table would be 371 inches below a Minden solum that formed (Table 4.2). This soil is in the weakly developed subset (Fig. 4.2). In the Winterset soil a saturation zone would be only 19 inches below the base of the solum that formed, and in the Haig and Edina soils the saturation zones could be 6, 3, and 21 inches above the bases of the sola that subsequently formed. These soils are in the more strongly developed subset (Fig. 4.2), and obviously are wetter than the more weakly developed soils.

This effect of wetness on soil development was pointed

TABLE 4.2. POSSIBLE HEIGHT OF SOIL COLUMN OF C HORIZON MATERIAL ABOVE PALEOSOL SATURATED UNDER CURRENT ANNUAL RAINFALL

Soil	Loess thickness*	Depth to base of solum*	Volume weight*	Volume soil solids*	Total porosity*	Annual rainfall	Calculated height saturated column above paleosol	Calculated depth to possible water table
	(in.)	*(in.)*		*(%)*	*(%)*	*(in.)*	*(in.)*	*(in.)*
Minden	468	39	1.23	48.3	51.7	30	58	410
Winterset	130	46	1.38	51.0	49.0	32	65	65
Haig	112	46†	1.37	53.0	47.0	34	72	40†
Haig	105	34†	1.39	53.9	46.1	34	74	31†
Edina	90	36†	1.40	54.6	45.4	34	75	15†

* Data from Ulrich, 1950.
† Base of solum would be 6, 3, and 21 inches below top of zone of saturation.

out long ago (Bray, 1934–1935), but this explanation was rejected by later workers in favor of: (1) relative effective age of weathering during and after loess deposition (Smith, 1942); (2) effective age of weathering plus particle-size fractionation of parent material (Hutton, 1947, 1951); and (3) change to finer texture of parent material as the loess thins (Hanna and Bidwell, 1955).

We shall examine these explanations, the "how" and "why," within our functional system of southwest Iowa. In regard to the influence of the particle size of the C horizon affecting the clay content of the B horizon, undoubtedly a relationship may be established (Hanna and Bidwell, 1955) because as the B horizon forms from the C, it inherits the clay of the preexisting C horizon. However, if the increments of change of clay content of both B and C horizons are examined from site to site, a different picture may be shown. For example, in the Humic Gley soils and Planosols in southwestern Iowa the clay contents of the C and B horizons for four successive sites are 23.6, 28.8, 30.7, and 31.6 percent, and 30.2, 40.1, 49.5, and 52.3 percent, respectively. The increments of change between C horizons in order are 5.2, 1.9, and 0.9 percent, but between B horizons in order are 9.9, 9.4, and 2.8 percent. For the 10 successive sites in the study in Kansas (Hanna and Bidwell, 1955), the same haphazard sets of data are determinable. Something besides the initial clay content of the C horizon must be involved. In a study in Illinois the range of mean particle size for one member of the soil sequence is greater than for the other four members (Smith, 1942).

In Illinois the relative effective ages were estimated from carbonate losses through leaching, both during and subsequent to loess deposition (Smith, 1942). Carbonate losses were estimated from measurements and calculated as equivalent to a layer of calcium carbonate 17.7 inches thick since the close of loess deposition and a layer 29.8 inches thick during loess deposition. These values were transposed to time units of 1.0 and 1.7. The average age of the upper 30 inches of a 300-inch loess deposit would be the weathering since deposition, or 1.0 time unit plus a half of the time required to deposit a tenth of the loess or $1.7/10 \times \frac{1}{2} = 0.1$, yielding a total of 1.1 time units. For a 30-inch loess deposit, the value would be 1.0 plus $1.7 \times \frac{1}{2} = 0.85$ or 1.85. Between these end members, others of the sequence were calculated and assigned age values. All of these values supposedly explained the sequential differences in the soils.

In Iowa a similar approach was taken (Hutton, 1947, 1951). Effective age of weathering was calculated by relating solum thickness to loess thickness multiplied by a constant for time of weathering during loess deposition ($T2$) plus the time of weathering since loess deposition ($T1$). Where loess is 600 inches thick and the solum is 30 inches thick, effective age would be $30/600\ T2 + T1$ or $T1 + T2/20$. Where the loess is 90 inches thick and the solum is 30 inches thick, effective age would be $T1 + T2/3$. The time of loess deposition was assumed to have been the same at all sites, that is, 600 inches were deposited at one place and 90 at another during the same time ($T2$). The effective ages calculated for the soils, Monona, Marshall, Sharpsburg, Grundy, and Seymour, in order are 0.050, 0.087, 0.180, 0.300, and 0.312 (Hutton, 1951). Recalculated to base one for Monona, the effective age values in order are 1.00, 1.74, 3.60, 6.00, and 6.24.

Then along came radiocarbon dating and the monkey wrench fell into the wheels. The assumption that the duration of loess deposition period ($T2$) was constant at all places is not supported by radiocarbon dates at the base of the loess (Table 2.3). The age of the base of the loess (Y) is progressively younger as the distance from the Missouri Valley source (X) increases as expressed by:

$$Y = 26{,}500 - 55X$$

Any influence by effective age of weathering must be foreshortened accordingly. From this equation we can calculate the duration of loess deposition at any distance from the source. We know that deposition ceased 14,000 years ago

(Table 2.4). Therefore, weathering during loess deposition ($T2$) equals Y from the last equation minus 14,000 years. The time of weathering since loess deposition ($T1$) is the same at all sites, or 14,000 years. Using the equation of effective age, values can be calculated for soils such as Monona, Marshall, Sharpsburg, Grundy, and Seymour (Table 4.1). Reducing all values to base 1.0 for Monona, the value for Sharpsburg is 1.05 and for Seymour is 1.02. Yet, our diagnostic soil property of clay content of B horizon shows 30.4 percent for Monona and 34.3 and 53.2 percent for Sharpsburg and Seymour. Obviously, the effective age ratios are invalid, and the idea of effective age does not adequately explain the difference in soil development on the summit landscapes of southern Iowa.

Now let us summarize what we know about the landscape model and the soil system. We can do this with functions.

(1) The loess thins with distance as:

$$Y = 1250.5 - 528.5 \log X$$

(2) The particle size measured by median diameter decreases with distance as:

$$Y = 1/(0.047 + 0.002X)$$

(3) The width of summits broadens with distance as:

$$Y = 0.019 + 0.0076X$$

(4) The local relief decreases with distance as:

$$Y = 1/(0.0034 + 0.0012X)$$

(5) The depth to more impermeable paleosols equals the loess thickness at any place and is expressed by (1).

(6) The depth to the deoxidized zone, a zone of former and

at some places present water saturation, decreases with distance as:

$$Y = 434.9 - 173.7 \log X$$

(7) The depth to the deoxidized zone decreases as loess thickness decreases in a bipartite relation separable at a loess thickness of 160 inches. Where loess is more than 160 inches thick, then

$$Y = 44.86 + 0.26X$$

Where loess is less than 160 inches thick, then

$$Y = 0.08 + 0.56X$$

(8) The age of the base of the loess decreases with distance as:

$$Y = 26{,}500 - 55X$$

(9) The age of the surface of the loess on uneroded summits is constant, being 14,000 years. Thus, from (8) and (9) where loess is 600 inches thick in Monona County, it was deposited during 15,000 years, from 14,000 to 29,-000 years ago. Where loess is 95 inches thick in Wayne County, it was deposited during 3,000 years, from 14,000 to 17,000 years ago. Manipulation of (8) and (9) rules out effective age as the cause of differential soil development on the loess.

(10) The amount of clay in the B horizons of soils increases as loess thickness decreases as expressed in a bipartite relation separable at a loess thickness of 160 inches. Where loess thickness is less, then

$$Y = 71.4 - 0.23X$$

Where loess thickness is greater, then

$$Y = 35.6 - 0.01X$$

The rate of increase in clay is 23 times greater where loess thickness (or depth to paleosol) progressively decreases 74 inches than where loess thickness progressively decreases 434 inches.

(11) The amount of clay in the B horizon of soils increases as the depth to the deoxidized zone decreases, as expressed in a bipartite relation separable at depth of 90 inches. Where the depth is less, then

$$Y = 69.1 - 0.37X$$

Where the depth is greater, then

$$Y = 36.9 - 0.03X$$

The rate of increase in clay is 12 times greater where the depth progressively decreases 40 inches than where the depth progressively decreases 120 inches. Within this analysis, the functions of (3), (4), (5), (6), (7), (10), and (11) point toward a more moist soil environment caused by subsurface zones of saturation perched on more impermeable paleosols at shallow depths. Such environment occurs where the rate of soil development is greatest. Conversely, where loess is thicker, paleosols deeper, and subsurface saturation zones deeper, a less moist soil environment would occur. This is where the rate of soil development is much less.

In other parts of Iowa and elsewhere, this basic model and system are usable.

We have previously discussed the complexity of landscapes within the loess country of Iowa (Chapter 3), and that interfluve summits do not descend from interstream divide summits along one continuous long slope. Instead, a staircase of levels like treads and risers is present. Such levels are distinctly visible if one looks into a watershed from a peripheral divide (Fig. 4.3). In the southern half of the state, the upper steps of the staircase generally are underlain by various paleosols that are on buried geomorphic surfaces. On the lower steps, the loess overlies buried erosion surfaces cut in till or other deposits (Chapter 3).

In northeast Iowa in the region of the Iowan erosion surface, the landscape also is laid out in the typical staircase pattern along interfluves. Most of the treads below the highest one have the erosion surface beneath loess and without an intervening paleosol. Commonly, a loess-mantled terrace along a stream will be the lowest tread overlain by eolian silt. This will be our model for formulation of principles about landscapes.

In viewing this multileveled terrain, one might expect to find different erosion levels, each with characteristic soils that differ from those on lower or higher levels. Such is not the case. In a local area, eolian deposition of loess may result in the homogenization of the soil landscape so that very similar soils are on all levels.

FIG. 4.3. View into drainage basin along North Turkey Creek in Adair County, Iowa, showing stepped topographic levels along interfluves.

In Tama County, Iowa, the landscape rises along interfluves in a series of stepped levels from Wolf Creek toward a divide summit a few miles to the south. The last increment of Wisconsin loess was deposited on all levels. On similar local topographic sites on all levels where postloess erosion is most precluded, the land surface dates from 14,000 years ago. In this area an alluvial terrace, 19 feet above Wolf Creek, has 4 feet of loess on it. The next higher level is part of the Iowan erosion surface that ascends to a height 130 feet above the creek and has 5 to 7 feet of loess on it. A second higher level ascends to a height of 155 feet above the creek and has 7 to 10 feet of loess on it. The divide is 196 feet above stream level and has 35 feet of loess. On these four steps the 4 feet of loess just beneath the land surface is the same last deposited increment of the loess.

What are the soils like on the four topographic levels? The Tama soil, characteristic of one formed under prairie, is on all of the surfaces. Regardless of level, the soils are almost as alike as peas in a pod. Examination of properties of the soil profiles shows this (Fig. 4.4). Clay, organic carbon, and organic phosphorus distributions in the vertical profiles are very similar. Recognizable dissimilarity appears only in the profile distributions of cation-exchange capacity and base saturation. The major difference is in the soil in the loess on the alluvial terrace. The underlying alluvium is permeable sand at least 40 feet thick. One would expect, then, a more intensive leaching of bases here in contrast to levels where the shallow substratum is more impermeable till.

The principle shown here is the formation of a local landscape with very similar soils that developed in a layer of similar sediment, regardless of topographic level, in the same period of time.

Drift Province

The drift province of Iowa has two major regions. One is the Cary drift of the Des Moines lobe and the other is the loess-mantled area of the Iowan erosion surface in northeast Iowa. The major landscape features of the Cary drift have already been discussed (Chapter 2) because of the intimate relation that exists between the kind of material deposited by glacier ice and the landscape that is left when the ice uncovers it. These land features are the end moraines, ground moraines, outwash plains, and valley trains that are parts of the morphostratigraphic units that comprise the Cary drift. Suffice it to say that the maximum age of any summit in this region cannot be greater than 13,000 years. This is the young-

FIG. 4.4. Properties of the Tama soil on four different landscape levels along Wolf Creek in Tama County, Iowa.

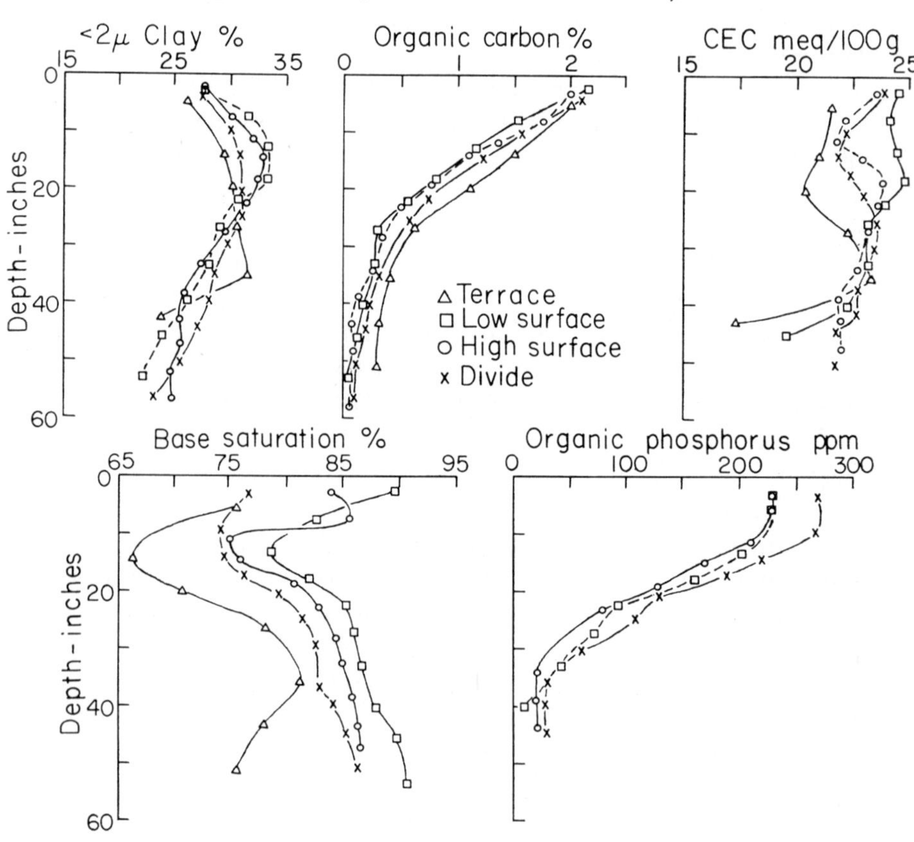

est time of uncovering of the surface by ice in Iowa (Table 2.7). But, as we shall see when we examine hillslopes, most summits on the Des Moines lobe are younger. Erosion on hillslopes has extended headward and upward so that even the summits have been narrowed and probably lowered. Consequently, most of the summits are much younger than 13,000 years.

The loam-mantled Iowa erosion surface is a different problem. Here, drainage nets are integrated throughout the region, and most summits change imperceptibly to hillslopes which descend to the valleys. Consequently, most summits may be the same age as most hillslopes, and can be exceedingly youthful. At the present time in our studies, the maximum determined age of a summit in this region is 12,700 years (I-2333). Unless proved otherwise in further studies, most of the land surface in this region equates in time to the land surface on the Cary drift. Both are less than 13,000 years old (Chapter 3).

HILLSLOPES

Hillslopes are a larger part of the land surface of Iowa, and are some of the most important landforms from the viewpoints of geomorphology and pedology. They may be formed by the reduction of a landscape by erosion, or they may be constructed by water, wind, or glacier ice. Regardless of the primary mode of origin, most if not all hillslopes in this state have been modified in very recent time by processes of erosion, sedimentation, mass movement, weathering, and soil formation.

As in our discussion of hill summits, we cannot possibly handle all hillslopes in Iowa, but we can set up certain models and systems which have wide regional application not only here but in other parts of the United States and abroad. We will present examples of these systems, and formulate principles that will lead to an understanding of similar landscapes. No hillslope system is completely understood unless soils are related to it. This is frequently overlooked in even the most sophisticated research studies on hillslope evolution.

Before we can examine our models and systems, a rudimentary jargon is needed as a basis of communication.

Geometric and Geomorphic Components of Hillslopes

Slope may be defined in at least two ways. Geometrically, it is the gradient at which a surface deviates from the horizontal. Geomorphically, slope is an inclined land surface of any part of the earth. The terms slope and hillslope are synonymous in this latter sense. Geometrically, any hillslope may be defined in space by three components: (1) its inclination to the horizontal plane, *gradient;* (2) its distribution along the direction of gradient, *slope length;* and (3) its distribution along the contour normal to the slope length, *slope width.*

Any hillslope may be straight or curved along length and width. Along slope length, if gradient is constant per unit of length, the slope is linear. If gradient increases or decreases per unit of length, the slope is curved. Upslope, if gradient increases per unit of length, the slope is concave, but if gradient decreases per unit of length, the slope is convex. Downslope, the reverse is true.

Along slope width, if slope length direction does not change, the slope is linear. If slope length directions change, the slope is curved laterally. Downward convergence of slope lengths yields slope width concavity, and downward divergence yields lateral convexity. The three-fold combination of linearity, convexity, and concavity along slope length and slope width forms nine basic slope geometries: (1) linear, convex, or concave slope width with linear slope length; (2) the same slope widths with convex slope length; and (3) the same slope widths with concave slope length.

The geometry of the hillslope in space can be measured

very simply. One need only measure unit lengths and gradients in an arranged pattern. However, if one wishes to dress up a study a bit, he may resort to other kinds of analytical measures (Savigear, 1956; Strahler, 1956).

Parts of hillslopes conform to the physiography of a landform. After stream incision, a valley is part of a drainage net that descends to and joins a larger trunk. Debris eroded from the valley slopes may be confined in a lower alluvial fill that is bounded on three sides, but part of the sediment may be carried from the tributary valley into the alluvial fill of the larger trunk. Thus, the tributary valley is open at one end and is an *open system*. Its sedimentologic record may be incomplete.

In a *closed system* an encircling bounding slope descends to a common depository of sediment, so that all eroded debris is trapped within the system. The sedimentologic record is complete. The slope descending to the depository is the *peripheral slope,* but where a closed system has been embayed, components of the open system may occur.

Closure of an open system on only three sides permits definition of geomorphic components of headslope, sideslope, and noseslope (Fig. 4.5). Within these geomorphic components, slope-profile components may be defined as *summit, shoulder, backslope, footslope,* and *toeslope* (Fig. 4.5). Discordance in gradient aids in delineating them. Inclination of the summit may differ distinctly from the hillslope which ascends to it. The shoulder commonly is the convexly rounded component between the summit and the backslope. The backslope usually is the linear part of the hillslope. The footslope is the concave part of the hillslope that welds the linear segments to lower terrain and is in part erosional and in part depositional. The toeslope commonly is formed on depositional debris that extends away from the base of the hillslope.

All slope-profile components do not necessarily occur on every hillslope, or each may be only a minor part. On a continuously curving slope the toe- and footslope may merge upward with the shoulder and summit. The backslope may be suppressed. A linear backslope may angularly join a summit and toeslope. The footslope and shoulder are suppressed and cartographically are only lines on a map.

Generally there is a specific fit of kinds of soils to the various components of hillslopes of open and closed systems. The fit is so distinct that it may be expressed in mathematical terms as we shall see.

The "when" of these systems can be determined by applying the principle of descendancy (Chapter 1). A hillslope

FIG. 4.5. Geomorphic components of hillslope. Slopes in an open system: *Headslope* is at the head of the valley, and slope lengths converge downward. *Sideslope* bounds the valley along the sides, and slope lengths generally are parallel. *Noseslope* is at the valleyward end of interfluve and slope lengths diverge downward. On slope profile, *summit* is upland surface and descent downslope successively crosses *shoulder, backslope, footslope,* and *toeslope*.

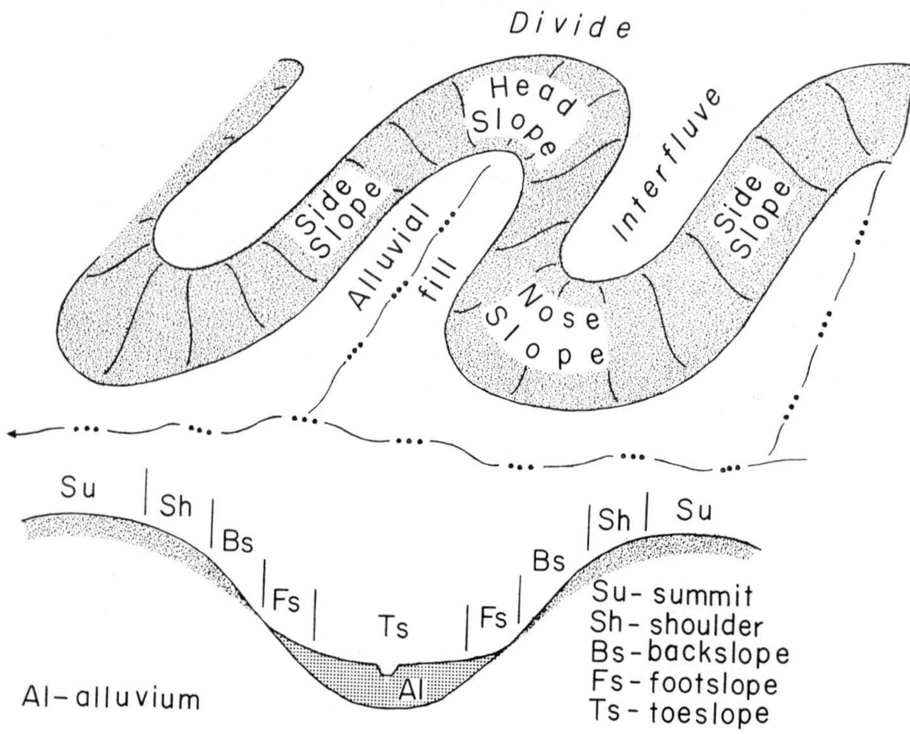

can be no older than the depositional debris that was derived from the hillslope and to which it descends. Numerous radiocarbon dates from sediments in open and closed systems in Iowa permit us to build absolute time into the framework of our systems (Table 4.3). All hillslopes in the state can be studied from the systems approach. The Des Moines drift lobe contains innumerable closed depressions, so the closed system can be used here. Some drainage nets have formed on the lobe, so the open system can also be used. In the remainder of Iowa, drainage nets are completely integrated so that open system analysis is the rule.

Open System

A few examples of open systems from Adair (Fig. 4.6) and Harrison counties will be explained, and the principles formulated have wide application throughout Iowa and elsewhere. In each case the possible sediment source for the alluvial fill in the valley is confined to the side- and headslopes in the open system. Localized sources of high sediment yield are lacking. Radiocarbon and historical dates at the base of and within the alluvial fills establish a chronology not only for sedimentation but also for hillslope erosion. Volumes of sediment in the fills permit an estimate of erosion on the bounding hillslopes.

How do we go about the study? Our first example of the open system (Ruhe and Daniels, 1965) is a small side valley in Adair County, Iowa, which is incised in Kansan glacial till that is overlain by 16 to 18 feet of Wisconsin loess on the divide above the headslope. The loess also forms the summits and shoulders above the sideslopes. Backslopes on the head, side, and noseslopes are on till. Foot- and toeslopes are on alluvium. From detailed drilling data, the side valley was V-shaped prior to valley filling (Fig. 4.7), but valley filling resulted in a U-shaped transverse profile. Transverse and longitudinal cross sections show the geometric fit of the allu-

TABLE 4.3. RADIOCARBON DATES FROM VALLEY ALLUVIUM AND BOG SEDIMENTS IN IOWA

Sample*	Date in years before present	Location	Notes
Valley alluvium			
W-701	< 250	Thompson Creek, Harrison County	Turton alluvium, box-elder wood
W-799	1,100 ± 170		Mullenix alluvium, walnut wood
W-699	1,800 ± 200		Mullenix alluvium, willow wood
W-702	2,020 ± 200		Hatcher alluvium, red-elm wood
I-2334	1,830 ± 100	Centerville, Appanoose County	Chariton alluvium, red-elm wood
I-1421	2,080 ± 115	Wolf Creek, Tama County	Terrace alluvium, American-elm wood
M-932	4,720 ± 250	Turin, Monona County	Gully fill, human bone
M-1071	6,080 ± 300		Gully fill, bison bone
W-235	6,800 ± 300	Cut 39, Pottawattamie County	Gully fill, organic carbon†
I-79	8,430 ± 250	Quimby, Cherokee County	Little Sioux alluvium, charcoal
W-700	11,120 ± 440	Thompsin Creek, Harrison County	Soetmelk alluvium, spruce wood
W-882	11,600 ± 200	Willow River, Harrison County	Soetmelk alluvium, spruce wood
W-881	14,300 ± 250		
I-2329	18,300 ± 310	4-Mile Creek, Tama County	Creek alluvium, organic carbon
Bog sediments			
I-1013	3,100 ± 130	Colo bog, Story County	Peat, 34–36 inches; grass-herb zone‡
I-1014	8,320 ± 275		Muck, 132–134 inches; conifer-hardwood zone, transition to grass
I-1015	13,775 ± 300		Peaty muck, 186–189 inches; conifer zone
I-1016§	2,365 ± 500	Jewell bog, Hamilton County	Peat, 24–26 inches; grass-herb zone
I-1417	9,570 ± 180		Peaty muck, 176–180 inches; conifer-hardwood transition zone
I-1017	10,226 ± 400		Muck, 210–212 inches; conifer-hardwood zone
I-1418	10,640 ± 270		Silts, 236–240 inches; conifer zone
I-1018	10,670 ± 400		Silts, 280–282 inches; conifer zone
I-1019	11,635 ± 400		Silts, 336–342 inches; spruce wood

TABLE 4.3. *(Continued)*

Sample*	Date in years before present	Location	Notes
I-1412§	3,170 ± 190	McCullock bog, Hancock County	Peat, 36–38 inches; grass-herb zone
W-551	6,570 ± 200	McCullock bog, Hancock County	Muck, 78–90 inches; grass zone; residue
W-554	6,580 ± 200	McCullock bog, Hancock County	Same, but humic-acid extract
W-549	8,170 ± 200	McCullock bog, Hancock County	Muck, 102–114 inches; hardwood-grass zone
W-553	8,110 ± 200	McCullock bog, Hancock County	Same, but humic-acid extract
I-1413	8,210 ± 260	McCullock bog, Hancock County	Muck, 135–137 inches; hardwood-grass zone
W-548	11,660 ± 250	McCullock bog, Hancock County	Muck, 150–156 inches; conifer zone
W-552	11,790 ± 250	McCullock bog, Hancock County	Same, but humic-acid extract
I-1414	14,500 ± 340	McCullock bog, Hancock County	Muck, 232–234 inches; conifer zone
I-1852§	2,830 ± 115	Woden bog, Hancock County	Muck, 25–28 inches; grass-herb zone
I-1853	5,390 ± 125	Woden bog, Hancock County	Silts, 96–102 inches; grass-herb zone
I-1415	7,050 ± 210	Woden bog, Hancock County	Muck, 291–294 inches; hardwood-grass zone
I-1854	7,770 ± 140	Woden bog, Hancock County	Muck, 318–321 inches; hardwood-grass zone
I-1855	9,300 ± 130	Woden bog, Hancock County	Silts, 330–334 inches; hardwood zone
I-1416	11,570 ± 330	Woden bog, Hancock County	Silts, 378–390 inches; conifer zone
I-1856	3,340 ± 110	Hebron bog, Kossuth County	Muck, 24–27 inches; grass-herb zone
I-1857	8,880 ± 140	Hebron bog, Kossuth County	Silts, 75–78 inches; conifer-hardwood zone
I-1858	27,900 + 1,100‖ − 1,000	Hebron bog, Kossuth County	Muck, 111–114 inches; conifer zone
I-1859	30,300 + 1,500‖ − 1,300	Hebron bog, Kossuth County	Silts, 144–148 inches; conifer zone
I-1860	2,930 ± 110	Sumner bog, Bremer County	Muck, 27–30 inches; grass-herb zone
I-1861	6,130 ± 120	Sumner bog, Bremer County	Muck, 54–57 inches; tree-grass zone
I-1862	11,880 ± 170	Sumner bog, Bremer County	Muck, 88–91 inches; conifer zone, larch wood

* Sample numbers are I for Isotopes, Inc.; M for University of Michigan; and W for U.S. Geological Survey, Washington, D.C.
† Strontium carbonate conversion.
‡ Pollen zones.
§ Composite of two cores.
‖ Erroneous values.

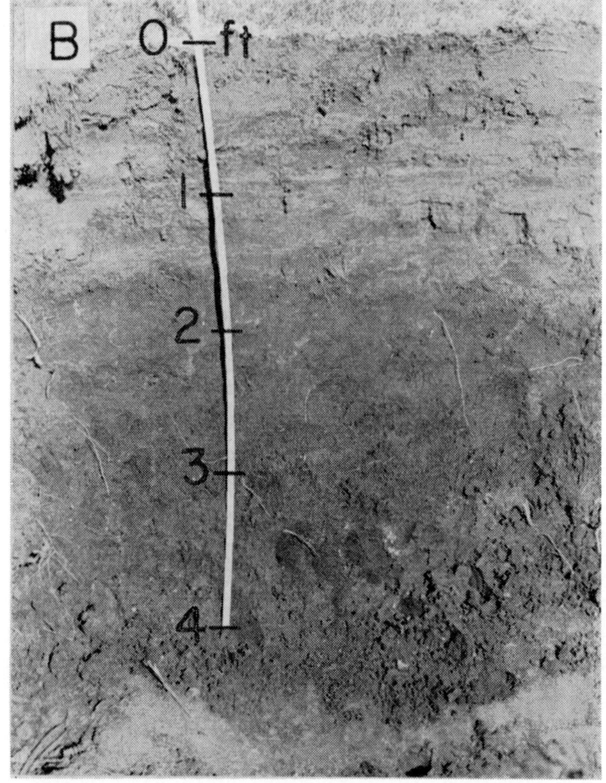

FIG. 4.6. (A) View into a side valley, an open system, in Adair County, Iowa. (B) Bedded postcultural sediment overlying dark A horizon in alluvial fill in side-valley bottom.

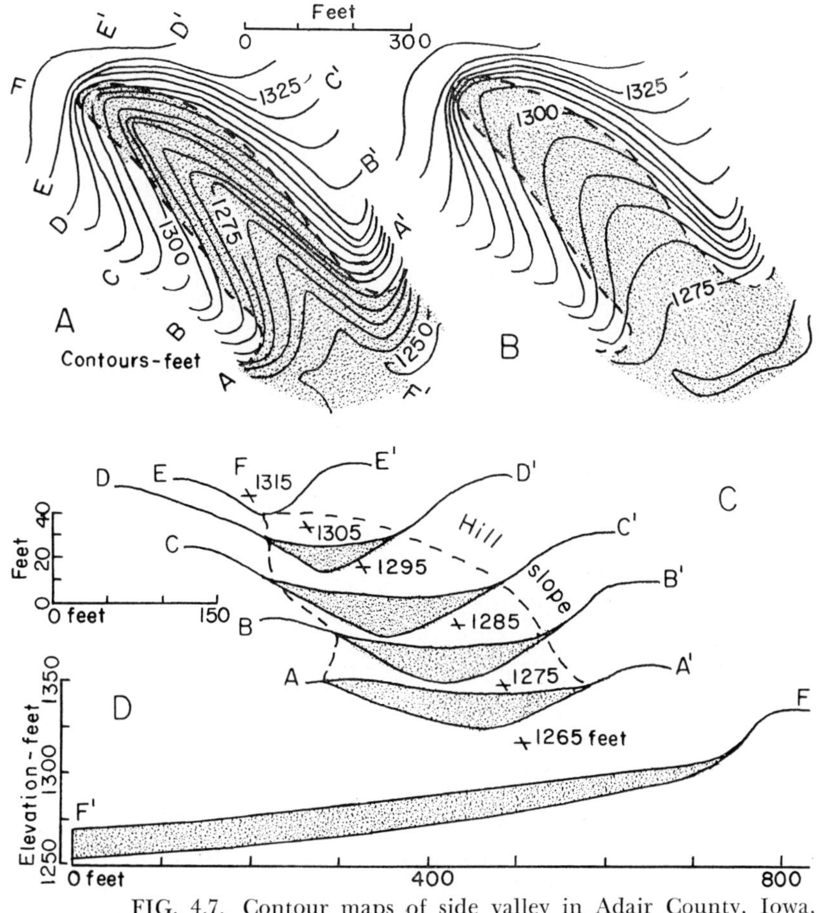

FIG. 4.7. Contour maps of side valley in Adair County, Iowa, after incision (A) and after alluvial filling (B). Stippled area delineates valley fill. Transverse (C) and longitudinal (D) cross sections of alluvial fill (lined areas). (From Ruhe and Daniels, 1965.)

vium in the valley. Along the axis the fill is uniformly 15 feet thick for 500 feet above the valley mouth, then thins gradually to 10 feet at station 660 and feathers out against the headslope.

The side valley is a self-contained landscape unit. All sediment had to be derived from adjoining side- and headslopes. Because the slopes are the source of sediment, their erosional age must be the same as that of the alluvial deposits to which they descend. The maximum age is equivalent to the age of the top of the fill. By radiocarbon dating, the base is 6,800 years old.

The alluvium has two distinct layers (Fig. 4.6). The upper bed is light colored and bedded, contains 14 percent sand and 1.8 percent organic carbon, and is a result of man's cultivation of the bounding slopes. The upper part of the lower bed is dark colored, has less than 2 percent sand to a depth of 4 feet, and contains 2.5 to 3.0 percent organic carbon. This layer is the A horizon of a buried soil that is presettlement in age. To a depth of 12.5 feet, sand content is less than 3 percent, indicating that the source of sediment was loess on the bounding valley slopes. Below 12.5 feet, sand content increases to 22 percent, indicating that till on the bounding slopes was the dominant sediment source. Sand content in the postsettlement sediment shows that both loess and till served as sources.

From the geometry of the fill, the total volume is 23,565 cubic yards. As the average thickness of postsettlement deposits is 6 inches, their volume is 1,394 cubic yards. The lower bed volume is 22,171 cubic yards. As the base of the fill is 6,800 years old and the area was settled 125 years ago, calculated rates of deposition are 3.3 cubic yards per year for the lower bed and 11.2 cubic yards per year for the upper bed.

The field data show: (1) a sediment source area in a confined physiographic system, and (2) a lack of localized sources on the bounding slopes such as gullies (Fig. 4.7). The only conclusion that can be reached is that each alluvial bed was probably derived uniformly from the bounding side- and headslopes. Now, the source area after valley incision was 191,754 square feet, but after deposition of the lower sediment, the source area was reduced to 75,262 square feet. Thus, 3.1 feet had to be stripped uniformly from the source area to form the lower fill, but only 6 inches were stripped to form the upper bed. Rates of slope reduction, then, were 0.006 and 0.06 inch per year, respectively.

A second example of an open system is a side valley that

descends to Turton Branch of Thompson Creek in Harrison County, Iowa (Ruhe and Daniels, 1965). The valley is incised entirely in loess, and its watershed contains 30.2 acres of which 3 acres are the convex summits and shoulders and 20.2 acres are backslope. Slope gradients on the slope components range from 0 to 2 and 7 to 35 percent, respectively.

The backslopes descend to an alluvial fill whose base is radiocarbon dated from a willow stump rooted in place 1,800 ± 200 years ago (W-699; see Table 4.3). A walnut log 3 feet above the base of the fill is 1,100 ± 170 years old (W-799) and 75 percent of the bed is above the log. Thus, the lower 25 percent of the bed is 1,100 to 1,800 years old and was deposited in 700 years. The alluvial fill, at other places, has a channel fill of alluvium inset in its upper part, and a box-elder stump rooted 2 feet above the base of the channel fill is 250 years old (W-701). Hence, the upper 75 percent of the alluvial fill in the side valley is 250 to 1,100 years old and was formed during 850 years. The area was settled about 115 years ago. Such detailed chronology permits a more accurate evaluation of sediment production and valley slope erosion.

The alluvial fill occupies 7 acres, and its volume, calculated from 8 transverse drilling traverses, is 149,222 cubic yards. The entire fill was deposited between 250 and 1,800 years ago at a rate of 96.3 cubic yards per year. As the lower quarter of the bed was deposited in 700 years, its rate of deposition was 53.6 cubic yards per year. The upper three quarters were deposited at a rate of 132.5 cubic yards per year for 850 years.

The volume of postsettlement sediment is 7,228 cubic yards. During the 115 years that it has accumulated, its rate has been 62.9 cubic yards per year.

Transposing the sediment volumes in time to the bounding head- and sideslopes shows that during postsettlement time 0.19 foot was stripped at a rate of 0.02 inch per year. During the time of 250 to 1,100 years ago, 3.15 feet were eroded at a rate of 0.04 inch per year, and during the time of 250 to 1,800 years ago, 3.98 feet were eroded at a rate of 0.03 inch

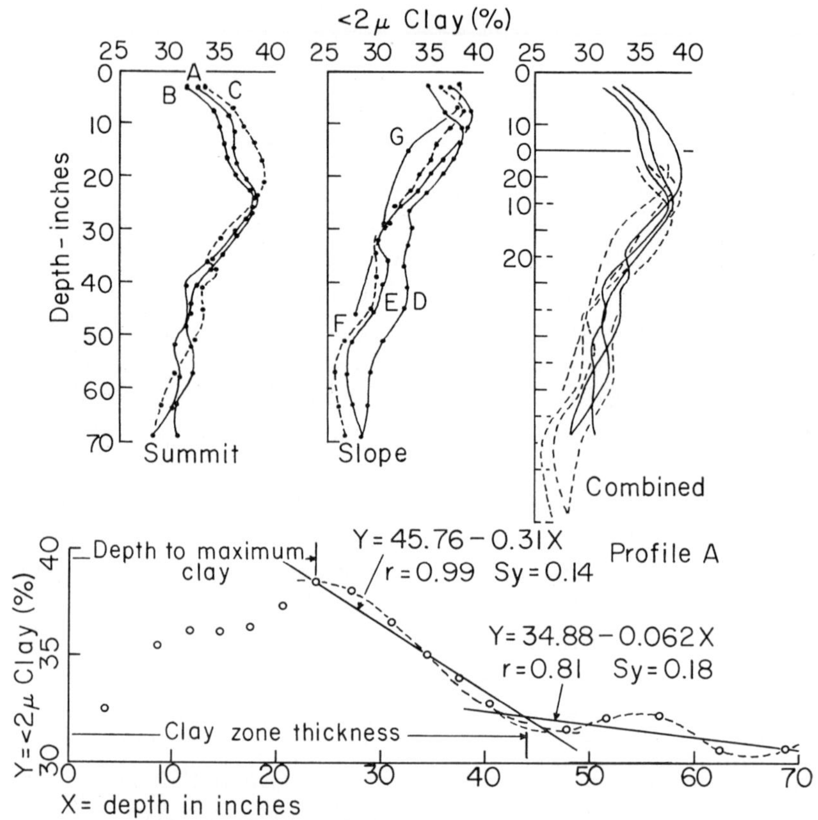

FIG. 4.8. Plot of clay distribution curves of soils on summit and hillslopes in Adair County, Iowa. Note two distinct families of curves. Combine two families by overlaying slope family on summit family and sliding slope family downward. Note similarity of all curves below clay bulges. Rotate clay curve *A* 90 degrees and mathematically fit trend lines to curve to determine clay zone thickness. Pick off depth to maximum clay.

per year. Again, these are minimum values because the amount of sediment removed through the open end of the system cannot be determined.

To analyze soil properties versus the landscape in the Adair area (Ruhe, Daniels, and Cady, 1967), it is first necessary to plot a specific property for each soil site on the landscape. We select the vertical clay distribution for each soil and plot it (Fig. 4.8). The curves of all hilltop sites are very similar, and the curves of all slope sites are similar. The two families of curves appear to differ. However, when the curves of the slope sites are superimposed on the curves of the hilltop sites and the slope family is moved downward on the hilltop family, the lower parts of the combined curves are very similar (Fig. 4.8). All of these soils have a common kind of clay distribution at some depth in their profiles. What are the depths of commonality?

The critical depths appear to be near the base of the bulges that show in the upper parts of the curves. These depths may be determined graphically by fitting linear segments to the curves and picking off the points where the changes in curvature at the base of the bulges are distinct. The base of the bulge may also be determined by mathematically fitting trend lines to the curves and noting the points of intercepts of the trend lines. An easy way to perform this latter analysis is to rotate the clay distribution curves 90 degrees and let the depth be X and the clay content be Y (Fig. 4.8). All values of X and Y, then, are positive and the arithmetic is simplified. By either method the base of the clay bulge, or the *clay zone thickness,* can be reasonably determined, and the agreement between values determined by each method is also reasonable (Table 4.4). Thus, one property, the clay zone thickness, can be determined for each soil and other properties also may be determined. Soil parameters may then be fitted to landscape parameters.

Clay zone thicknesses of soils, determined by graphic fit, decrease exponentially from 46 to 43 inches on the level summit to 39, 31, 24, 24, and 22 inches on respective slope gradients of 1, 2, 6, 8, and 9 percent. On the level summit, depths to maximum clay in the clay zone (Fig. 4.8) are 26 and 23.5 inches. As gradient increases downslope, the depths to maximum clay exponentially decrease to 20.5, 13.5, 7.5, 7.5, and 5.0 on the gradients of 1 to 9 percent, respectively. Note the relations of other soil properties to slope gradients (Table 4.4).

The depth to the maximum clay and the clay zone thickness versus slope gradient are not fitted by ordinary empirical curves. Instead, these relations are better expressed by the

TABLE 4.4. SOIL PROPERTIES FROM SUMMIT TO SHOULDER AND BACKSLOPE IN ADAIR COUNTY, IOWA

Soil	Slope gradient	Depth to maximum clay	Clay zone thickness		Weighted clay in zone	Depth to 1% organic carbon	Weighted organic carbon of > 1% content	Depth to pH 6.0	Depth to base saturation 80%
	(%)	(in.)	(in.)		(%)	(in.)	(%)	(in.)	(in.)
A	0	23.5	46	(44)*	35.1	22	1.83	39	46
B	0	26.0	43	(41)	34.7	21	1.94	43	28
C	1	20.5	39	(37)	35.1	19	1.71	48	26
D	2	13.5	31	(31)	34.7	12	1.60	35	25
E	6	7.5	24	(26)	34.4	15	1.74	18	21
F	8	7.5	24	(25)	33.8	12	1.50	15	15
G	9	5.0	22	(23)	32.7	9	1.32	9	9

* 46 by graphic fit to clay curve and (44) by mathematical fit to clay curve; see text; note reasonable agreement of values Fig. 4.9.

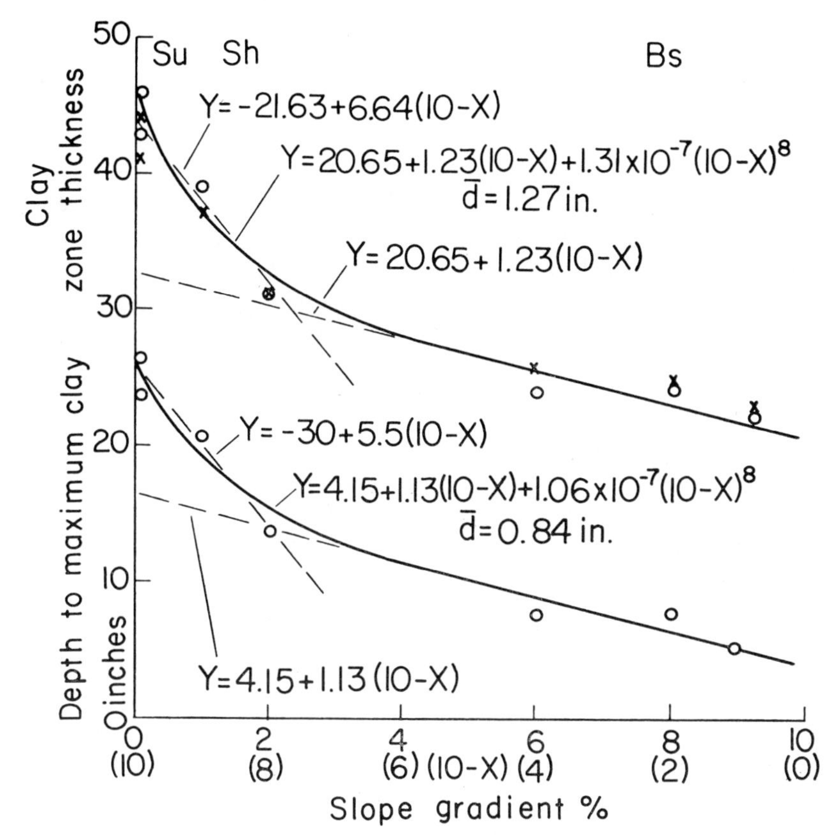

FIG. 4.9. Relationship of depth to maximum clay in the soil profiles and clay zone thickness to slope gradient in Adair County, Iowa. The set function is $Y = a + b_1X + b_2X^n$ and is composed of two subsets of linear functions of $Y = a + bX$. One subset is the shoulder and backslope *(Bs)*. Points: *o*—determined by graphic fit, *x*—by mathematical fit; see text.

general equation: $Y = a + b_1X + bX^n$ (Fig. 4.9) which is the set of two subsets of linear functions of $Y = a + bX$. One subset relates the properties on the summit and shoulder, and the other subset relates the properties on the shoulder and backslope. The discordance of subsets marks the lateral joining of the stable summit and the instable hillslope. If equations were calculated for other soil and landscape relations, discordance of subset functions also would be evident.

In the Harrison County study, all soils (Monona and associates) on all hillslope components are weakly developed, and B horizons are marked only by color and weak structure (Daniels and Jordan, 1966). However, A horizons are distinct, and certain physical and chemical properties of the soils are systematically related in the hillslope sequence. In fact, mathematical analysis of some of the data shows certain relationships that could not be observed in the field.

Very fine sand (0.10–0.05 mm) in all horizons of all profiles is generally 2 to 5 percent, but the distribution is not random. In the A horizons of < 12-inch depth and B horizons of 12- to 22-inch depth, very fine sand Y increases linearly with slope gradient X and can be expressed respectively as:

$$Y = 2.26 + 0.063X$$

and

$$Y = 2.23 + 0.067X$$

However, in the underlying C horizons there is no distinct relationship between sand content and slope gradient. Weathering cannot be responsible for this systematic organization of coarser particle sizes in the A and B horizons. Instead, slope wash probably has affected particle-size properties in difficultly identifiable surficial sediments on the hillslope. Coarser particle sizes have been left preferentially on steeper slope segments and prior to soil formation. The A and B horizons inherit these sedimentary properties.

The sedimentation effect is substantiated by examining coarse/fine silt ($50–20\mu/20–2\mu$) ratios in the A horizon Y versus two parameters, slope gradient in percent $X1$ and distance in feet from the summit $X2$ in multiple regression analysis. These relations are expressed as:

$$Y = 1.31 + 0.023X1 + 0.009X2$$

Now, if we resort to statistical testing, the silt ratio is noted to be equally dependent upon slope gradient and distance from the summit as shown by the sums of squares: R^2 (Y on $X1$) = 0.677 and R^2 (Y on $X2$) = 0.666. Slope gradient in its control of runoff, runin, and runon may be in part effective in differential weathering of silt, so that silt ratios systematically fit hillslope parameters. However, slope gradient and distance from summit are parameters also involved in sediment transport. Certainly, the latter is one not necessarily related to weathering at all.

Mathematical functions relate the properties of soil to hillslope parameters in both the Adair and Harrison studies. Yet, the relief in the former area is only 20 to 30 feet, whereas relief is 50 to 60 feet in the latter area. Slope gradients are less than 9 percent in the Adair case but as steep as 23 percent on the Harrison landscape. Common to each area, however, is the angular joining of a younger instable hillslope with an older stable upland summit. In the Adair model the hillslopes are less than 6,800 years, and the summit is 14,000 years old. In the Harrison model the hillslopes are less than 1,800 years, but the summit may be 14,000 years old. Within this framework of time, erosion with sedimentation and weathering affects the formation of soils on the landscape. Our mathematical functions express the systematic relations of soil properties on the landscape that was formed by the lateral mergence of two geomorphic surfaces of different ages.

This resorting to mathematics should not cause one to head for the door because, in reality, we have organized bundles of hillslope and soil measurements into one package. The equations are not very complicated, and they arrange and simplify all of the things that we want to look at in the system. These models can be used in other parts of Iowa where open landscape systems are present.

We can now turn to our closed system.

Closed System

Examples of slope reduction in a closed system are the Colo and Jewell bogs on the Des Moines lobe in Story and

Hamilton counties, Iowa (Walker, 1966). Peripheral hillslopes descend to a common depository of sediments that alternately have high contents of organic and mineral material (Figs. 4.10, 4.11). Mineral sediments in the bog were derived from the bounding hillslopes in glacial till. Now, bulk densities differ between surficial sediments on till, bog mineral sediments, and bog organic sediments. Average values are 1.22, 0.6, and 0.2 gm/cc, respectively. Consequently, if we want to compare the bog and source sediments, we must correct the bulk densities of the bog sediments so that they are equivalent to the source sediments. The correction is made

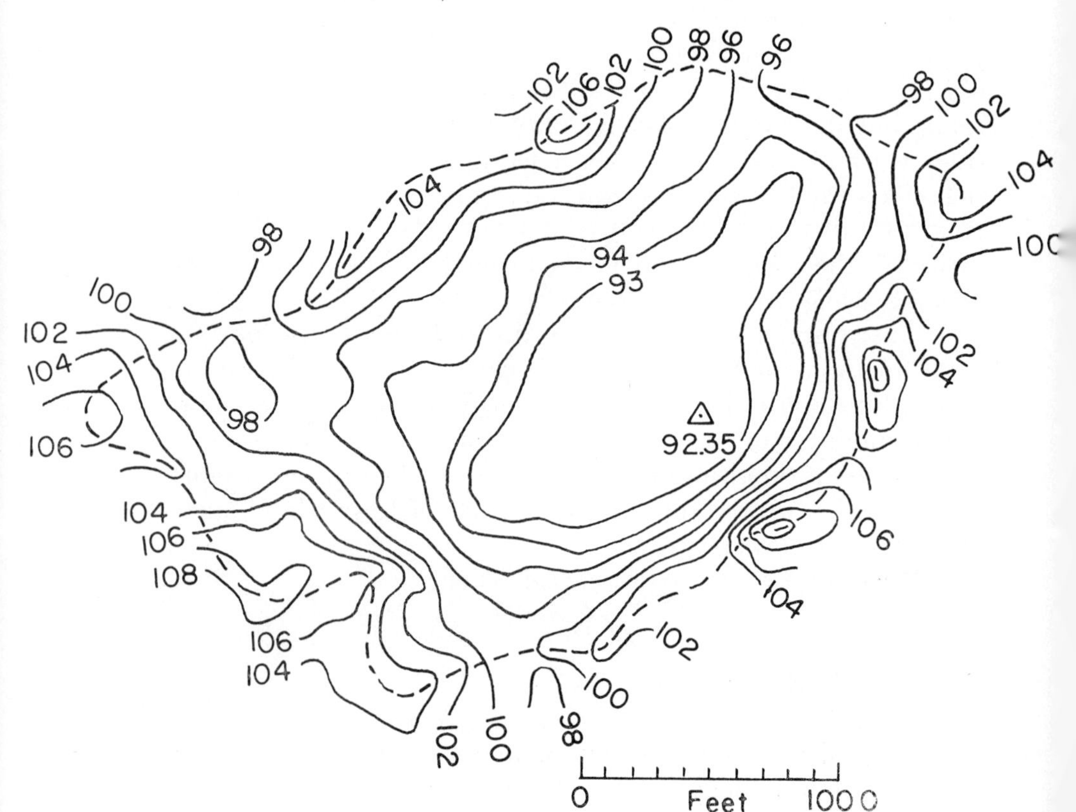

FIG. 4.10. Topographic map of Colo bog area showing peripheral slopes descending to a central depression. The divide is marked by a broken line. Elevation in feet above local datum. (From Walker, 1966.)

by multiplying the volume of a given bog sediment by the ratio of its bulk density to the bulk density of the source sediment. These results are the corrected volumes for bog sediments.

Using the same approach as in our open system of relating volume to the possible source area, we can calculate the erosion on the hillslopes around the bog. Thickness of the layer eroded equals the corrected volume of bog sediment divided by the area of the source. Around the Colo bog 0.2 foot was stripped from the hillslope during formation of the upper organic sediment (muck) in the bog, and 1.7 feet were eroded to form the upper mineral sediment (silt) in the bog (Table 4.5). Calculated values for the other sediments in the

FIG. 4.11. Exploded radial sections showing stratigraphy of Colo bog. Radiocarbon horizons are located. (From Walker, 1966.)

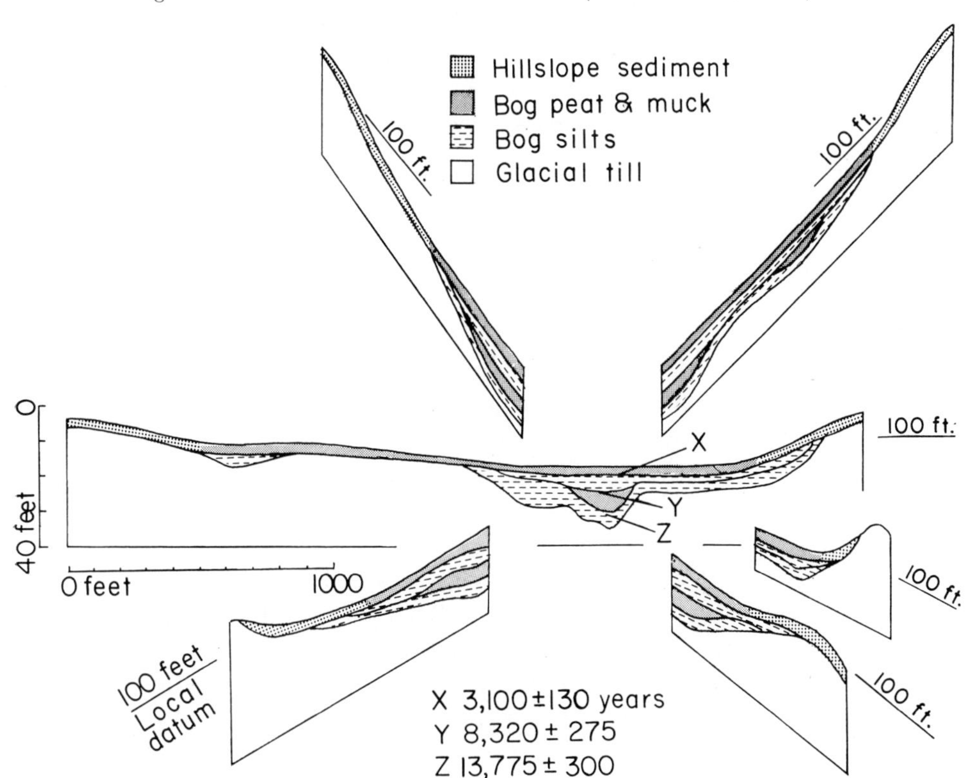

TABLE 4.5. HILLSLOPE REDUCTION ADJACENT TO COLO AND JEWELL BOGS IN CENTRAL IOWA

Bog stratum	Hillslope thickness	Erosion rate
	(ft.)	*(in./1000 years)*
Colo bog		
Upper muck	0.2	0.72
Upper silt	1.7	4.08
Lower muck	0.03	0.07
Lower silt	0.9	5.40
Total	2.8	2.57 (average)
Jewell bog		
Upper muck	0.2	1.08
Upper silt	5.3	8.76
Lower muck	0.5	6.00
Lower silt	0.6	7.20
Total	6.6	5.76 (average)

Source: Walker, 1966.

Colo and Jewell bogs are also given.

We can place the erosion and deposition systems within a time sequence by referring to our built-in isotopic chronometer. Radiocarbon dates for the Colo bog are 3,100 years at the base of the upper muck, 8,320 years at the top of the lower muck, and 13,775 years below the base of the lower muck (Fig. 4.11; Table 4.3). For comparable stratigraphic positions in the Jewell bog, the dates are 2,365, 9,570, and 11,635 years, respectively. Calculated slope reduction rates are 0.72 and 1.08 inches per 1,000 years for the upper muck zone and 4.08 and 8.76 inches per 1,000 years for the upper silt zone for the Colo and Jewell areas, respectively. The erosion rates differ between the Colo and Jewell hillslopes because of different slope gradients. At Colo the range in gradients is 2 to 3 percent, whereas at Jewell the gradients are 3 to 8 percent. The erosion was more intense on the steeper slopes during the same period of time.

Comparison of the closed system data with the open system data of the Adair and Harrison County areas also shows the effect of slope gradient on sediment production and erosion. Converting to inch-per-1000-year basis, the postsettlement rate in the Adair side valley is 60 inches per 1,000 years, and presettlement rate is 6 inches per 1,000 years. Slope gradients are 9 to 47 percent. The postsettlement rate in the

Thompson Creek side valley is 20 inches per 1,000 years. Rate for the period of 250 to 1,100 years ago is 40 inches per 1,000 years, and the rate for the period of 250 to 1,800 years ago is 30 inches per 1,000 years. Slope gradients are 7 to 35 percent. However, a material factor is also involved. The closed system slopes are on glacial till; the Adair open system slopes are on loess and till; and the open system slopes in Harrison County are entirely on loess.

Is there some kind of system involved on the hillslopes around closed depressions, and how do we examine it? First, we survey a transect from the summit to the toeslope and into the bog. Field measurements are made; samples are collected; and laboratory analyses are performed. Systematic changes are noted in thickness of sediment, content of gravel and clay particles, and geometric-mean particle size (Walker, 1966; Fig. 4.12). Thickness of surficial sediments increases. The gravel content decreases, and clay content increases so that the geometric-mean particle size decreases across the hillslope components. Wordwise, the analysis may be expanded by spelling out the detailed data along the transect of the hillslope profile.

However, the story can be simplified by relating a property to the hillslope transect, fitting curves to the data, and developing the empirical equations that express the relation. The thickness of surficial sediments on the hillslope is expressed by:

$$Y = 1.41 - 0.91X + 0.49X^2 - 0.034X^3$$

where Y is the thickness of sediment in feet and X is the distance from the summit down across the hillslope (Fig. 4.13). The distribution of gravel (2 to 2000μ) and clay ($< 2\mu$) in the surficial sediments is expressed respectively by:

$$Y_1 = 33.9 + 2.81X - 0.7X^2$$
$$Y_2 = 35.7 - 2.91X + 0.46X^2$$

where Y_1 is the weighted mean in percent of gravel (62 to 2000μ size) and Y_2 is the weighted mean in percent of the clay

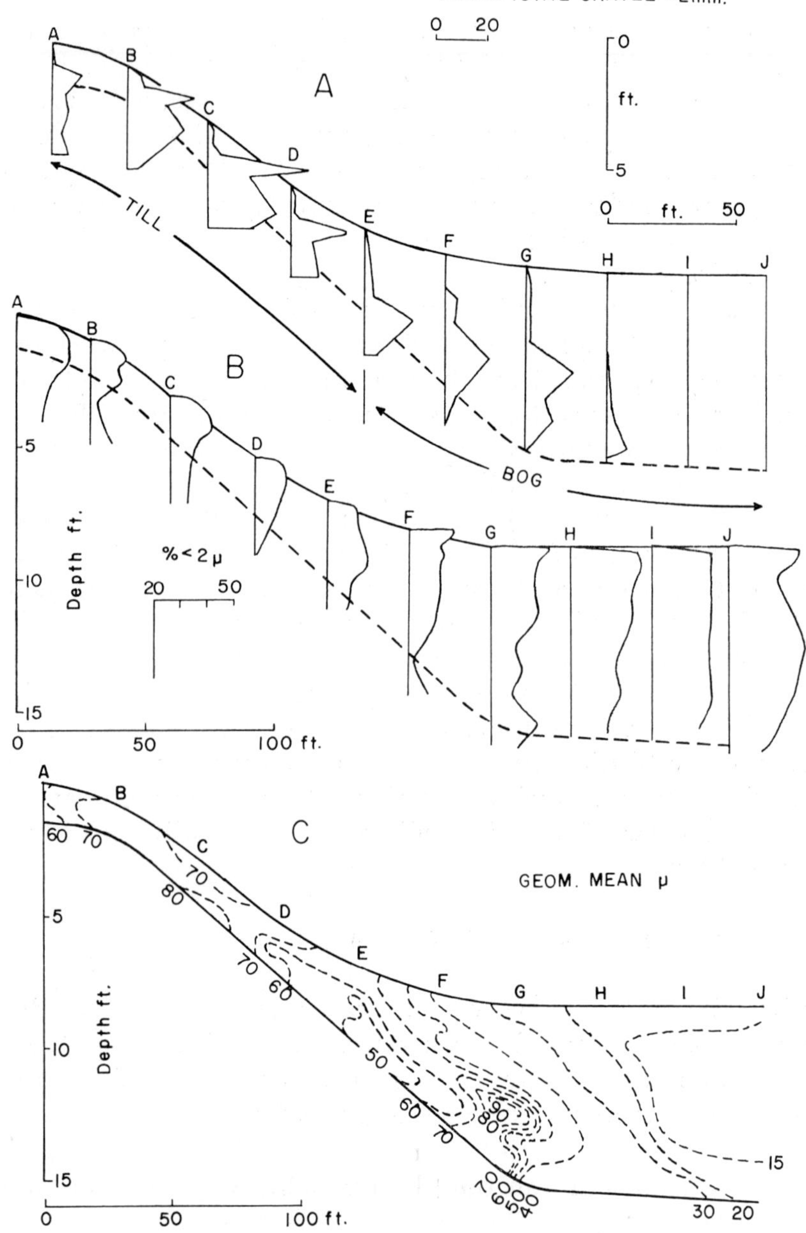

FIG. 4.12. Systematic change of thickness of surficial sediment and particle size related to hillslope around Jewell bog on Cary till in Hamilton County, Iowa. (From Walker, 1966.)

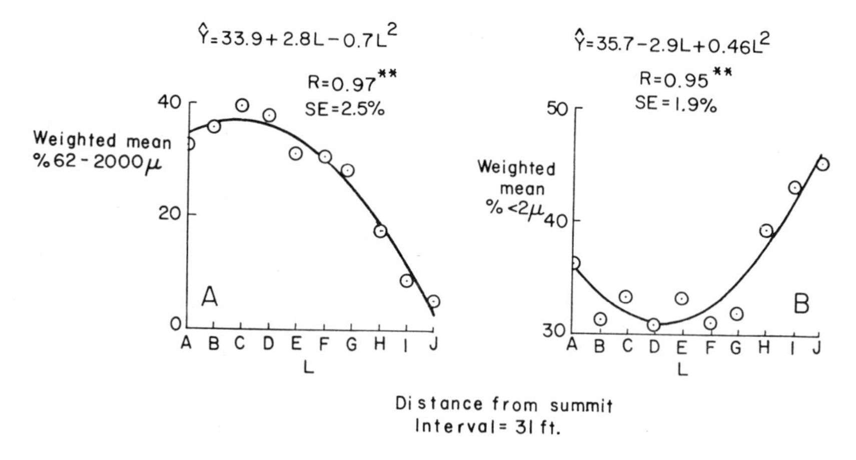

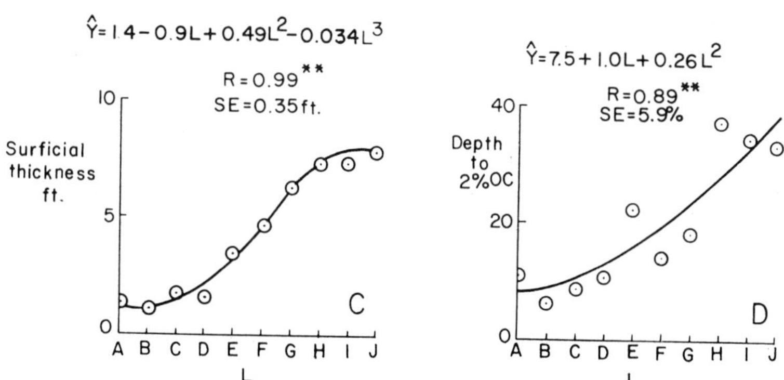

FIG. 4.13. Curve fitting and empirical equations expressing the relationship of selected properties to a hillslope on the Cary till around the Jewell bog in central Iowa. (From Walker, 1966.)

(< 2μ size) and X is the distance from the summit down the hillslope.

These data show that the hillslope is an erosional feature upon which sediment has been deposited. Not only does thickness become greater in a direction downslope, but there is distinct sorting of the sediment along the same direction. These phenomena can only be the result of deposition and sorting by water.

Some properties of the soils can be introduced in our system by analyzing organic carbon and calcium carbonate distributions (Fig. 4.14). These bring in weathering and soil

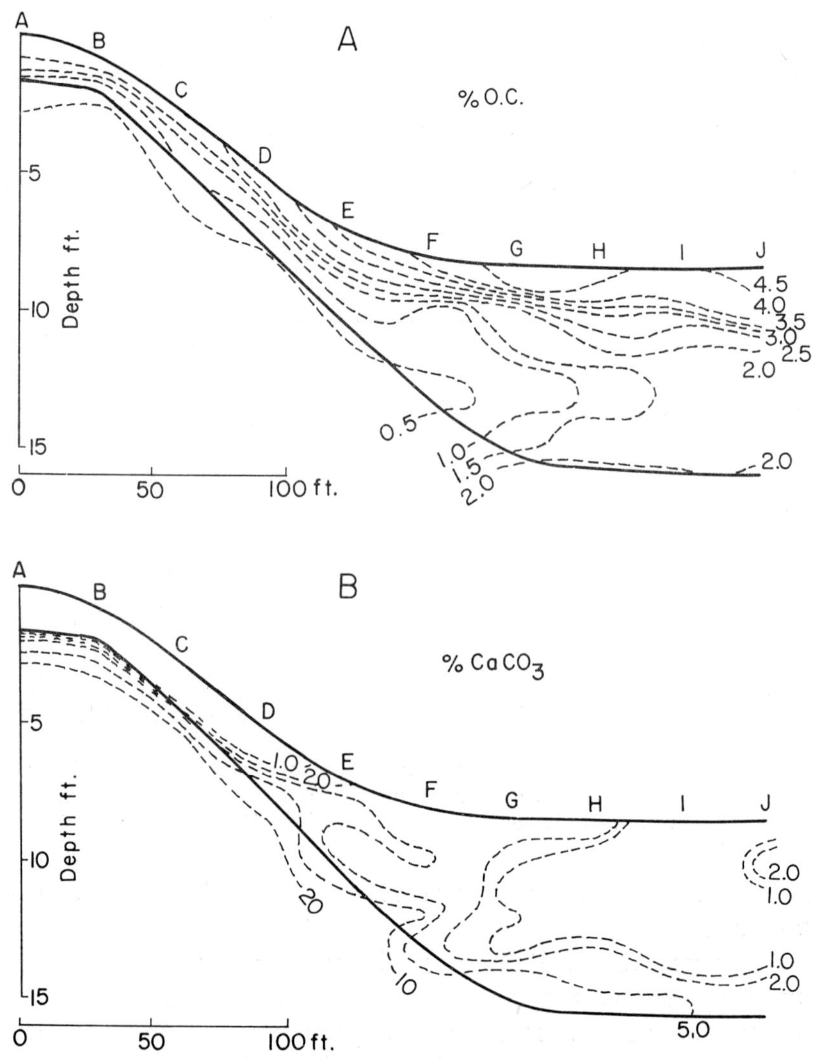

FIG. 4.14. Organic carbon and calcium carbonate distributions in the soils on a hillslope transect at the Jewell bog in central Iowa. (From Walker, 1966.)

formation on the hillslope. Organic carbon progressively increases from the summit, down across the hillslope, to the toeslope. Expressed as a function of depth to < 2 percent organic carbon (M), the relationship is:

$$M = 7.49 + 1.0X + 0.26X^2$$

Depth to carbonate progressively decreases to the footslope component.

At this point in our examination of the hillslope, we have determined that erosion has been active on it, that sediment has been deposited on components of it, and that the sediment has been selectively sorted from the summit to the toeslope and into the bog. In addition, during weathering and soil formation, organic carbon has accumulated preferentially toward the more poorly drained or wetter part of the landscape. Carbonates have been selectively leached to greater depths toward the better drained part of the landscape that is higher on the hillslope.

When did all of these things happen? Radiocarbon dates from key horizons in the Jewell bog give these answers. The lower peat in the bog was buried by upper silts 9,570 $\pm$ 180 years ago (I-1417; see Table 4.3). The upper peat at the bog surface began to form 2,365 $\pm$ 500 years ago (I-1016). Between these times and during the period of 7,200 years, a layer 5.3 feet thick was eroded from the adjacent hillslope at a rate of almost 9 inches per 1,000 years (Table 4.5). This was the time of instability of the hillslope, of erosion, of sedimentation with selective sorting down gradient, and of bog filling by mineral sediment.

During the past 2,365 years the hillslope has been more stable as only 0.2 foot has been eroded at a rate of only 1 inch per 1,000 years. This is the time of soil formation and weathering on the hillslope. Consequently, the soils on the hillslope are comparatively young—only a few thousand years old. Very similar histories are involved on hillslopes on the Des Moines lobe northward to the Minnesota state line. At Colo bog in Story County, hillslopes were stable during the last 3,100 years (I-1013), but subjected to more intensive erosion between 3,100 and 8,320 years ago (I-1014; see Tables 4.3, 4.5). At McCulloch bog in Hancock County, similar times were 3,200 (I-1412) and 8,200 years ago (I-1413; see Table 4.3). At Woden bog in Hancock County, the times were 2,800 (I-1852) and 7,700 years ago (I-1854), and at Hebron bog in Kossuth County they were 3,300 (I-1856) and 8,800 years ago (I-1857; see Table 4.3).

These bog areas are only five of twenty-three that were studied and which are located throughout the region of 12,300 square miles of the Des Moines lobe in Iowa (Walker, 1966). This also is the region of the Clarion-Nicollet-Webster soil association (Oschwald *et al.,* 1965). The similarities between the sample areas permit the formulation of a generalized landscape model and related soil system (Fig. 4.15). The Clarion soil commonly is on summits and shoulders with the Nicollet

FIG. 4.15. Generalized model of the Clarion-Webster soil system on the hillslopes on the Des Moines lobe in north central Iowa. Distribution of properties shown by arrow. (Modified from Walker, 1966.)

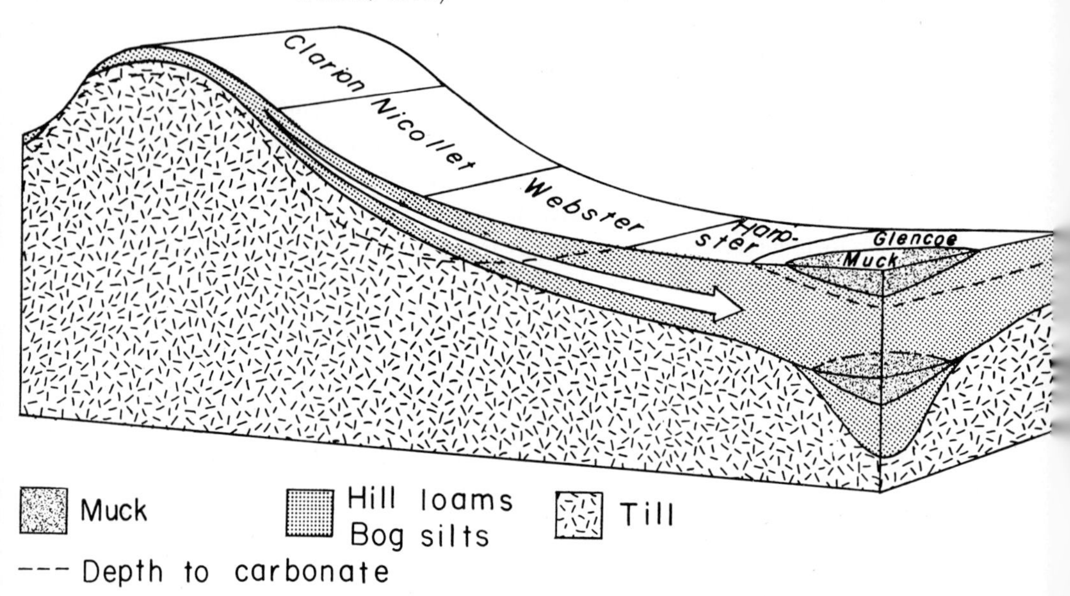

soil on the backslope and Webster soil or other associates on the foot- and toeslope. Organic soils such as peats and muck occupy the toeslope. These soils in their layout on the hillslope inherit the erosional and depositional characteristics that are so readily expressed by mathematical functions. Even their properties that form through pedogenic processes conform to functional systems.

From the geomorphic point of view, we have detailed the history of the upper silts and peat on the Des Moines lobe. There is also a buried lower pair of such sediments in the bogs. The radiocarbon dates (Table 4.3) for our five sample areas bracket the lower pair of sediments from about 13,000 to 7,000 years ago. Exclude the lower dates from Hebron bog which are old because of contamination (Part II). Erosion on hillslopes was comparable in intensity during the time that the lower mineral silts were being deposited in the bogs. Erosion rates were 5.4 and 7.2 inches per 1,000 years at the Colo and Jewell bogs (Table 4.5). Stability followed on the hillslopes as the highly organic lower peats and peaty mucks formed in the bogs. The upper episode followed.

This double nature region-wide pattern occurs throughout the Des Moines lobe. In all bogs studied to date, the two pairs of sediments are easily identified. Consequently, we can recognize two major erosion and deposition episodes in Recent time. The Colo cycle happened 7,000 to 13,000 years ago, and the Jewell cycle operated 3,000 to 7,000 years ago (pl. 1, legend).

As we have seen, much can be learned about hillslopes when studied from the approach of the closed system. We shall have to return later to bring the environment into the system. Before we do, let us look briefly at valleys in Iowa.

VALLEYS

Our interest in valleys of Iowa will be restricted to the landscapes of the floodplains and terraces. It will be almost impossible to cover all valleys, so the discussion will be confined to a reasonable number of models that exemplify regions. Our approach will be to return downstairs in the Pleistocene and work upward.

Establishment of the Drainage Net

The major valleys of southern Iowa (pl. 1) were formed during the late Yarmouth interglacial episode. How do we know that this was the time? By relating the Loveland loess to the land surface on which it was deposited. Recall that the Loveland loess is Illinoian age (Chapter 3), and that Kansan drift was deposited all over the southern half of the region (pl. 1). Further, recall that Yarmouth paleosols formed and were buried by the Loveland loess.

In the Rock Island Railroad cuts from Atlantic, Cass County, to Hancock, Pottawattamie County, the Yarmouth surface is marked by a paleosol at the top of a lake bed that overlies Kansan till. Loveland loess buries the paleosol and lake clay at many places. In 14 cuts through a distance of 13 miles, the top of the paleosol and lake bed has an arithmetic mean elevation of 1,286.5 feet. The standard deviation from the mean is 8.2 feet. The base of the lake bed is at an arithmetic mean elevation of 1,271.5 feet with a standard deviation of 4.8 feet. In ordinary words, these measurements show that the lake bed is an essentially horizontal key stratum throughout the region.

This traverse crosses primary streams, such as the Nishnabotna rivers and Keg Creek, and tributary streams, such as the Silver creeks and Walnut and Indian creeks, at right angles. The base of the Loveland loess is at elevations of 1,180 to 1,290 feet. This lower elevation is 92 to 106 feet below the mean elevation of the Yarmouth lake bed. Obviously, the Yarmouth level had to be incised by valleys before Loveland loess could be deposited below the mean Yarmouth level. The incision had to be after deposition of the Yarmouth lake bed and before the Illinoian deposition of Loveland loess, or in late Yarmouth time.

Where the Loveland loess is at elevations below the late Yarmouth datum level, the land surface buried by the loess descends toward a present main or tributary stream. Thus, an ancestral valley existed where the present stream is. Specifically, major streams such as Keg Creek and the West and East Nishnabotna rivers and large tributary streams such as Middle Silver, East Silver, Graybill, Walnut, and Silver creeks occupy valleys that originated in late Yarmouth time. The divides between the main stream and between the main and tributary streams also formed at that time. Stream cutting and valley filling have been going on in the same general places ever since.

Another line of evidence of the antiquity of the drainage net of southern Iowa is the three-dimensional distribution of the Late Sangamon erosion surface and its related depositional surface (Chapter 3). Universally, this surface descends from an interstream divide toward a stream valley. This descent can be in any direction. The present valleys are in the same general places that they were in Late Sangamon time.

In northwest and northeast Iowa, the various levels of the Iowan erosion surface descend to valleys, and also in any direction within the regions (Chapter 3). Consequently, the valleys had to exist 29,000 to 14,000 years ago. As previously explained, the drainage net on the Tazewell drift surface in northwest Iowa had to form 20,000 to 14,000 years ago, and the valleys in the Cary drift region in north central Iowa formed since 13,000 years ago (Figs. 3.12, 3.13).

Terraces and Benches

A *terrace* is a valley-contained feature that forms a level which is bordered on one side by a higher recognizable valley wall and on the other side by a scarp that descends to some lower level. In our usage, a terrace is composed of alluvial deposits. A *bench* is a planed feature physiographically similar in form to a terrace but which is an erosion surface in the valley (Tator, 1953). Terraces and benches may occur in a valley like a staircase from the floodplain to the bounding valley wall.

Many complications may be involved and identification and correlation of these features within a valley or from valley to valley may be very complex (Johnson, 1944; Frye and Leonard, 1954; Howard, 1959). A good example of this complexity is the Des Moines River Valley. This stream heads on the Cary drift of the Des Moines lobe in Minnesota. It extends southward into Iowa more or less along the axis of the lobe to the city of Des Moines where it leaves the Cary drift region (pl. 1). The river then crosses the loess-mantled Kansan drift region to the extreme southeast corner of Iowa where it joins the Mississippi River. The Des Moines is the largest tributary of the Mississippi in the state, and has a drainage basin of 14,467 square miles.

The valley is crossed by U.S. Highway 30 in Boone County in central Iowa and within the Bemis morainal system of the Cary drift. Here three distinct terraces stand above the floodplains and presumably are Cary and postglacial in age (Ruhe, 1965). Downstream in the big bend of the river near Keosauqua in Van Buren County, terrace levels stand 10, 15, 25, 50, 75, 90, 120, and 145 feet above the stream (Lees, 1916). What is the fit of the three levels in Boone County to the eight levels near Keosauqua? The answer is unknown. The details have never been studied. In fact, open season is in effect on terraces in Iowa. Little systematic investigation has been made of any terrace system in any valley.

However, a natural breakdown exists in the Des Moines Valley system. On the Cary drift lobe, all terraces are free of a loess mantle. In the Kansan area, lower terraces do not have loess caps, but the higher ones do. Immediately, a time separation can be made. Those terraces free of loess must be less than 14,000 years old, whereas those terraces that have a loess mantle must be older than 14,000 years. Why? Because we know that loess deposition stopped about 14,000 years ago (Table 2.4).

Loess-mantled terraces may have some peculiarities that are not shared by others in Iowa. This is particularly true in the western part of the state where loess is thick. There the thickness of loess on the terrace may be the same as the thickness of loess on the adjacent upland. The physiographic form, that is, a level within the valley, is preserved, but the sediment standing above the floodplain may be entirely loess and not alluvium. The terrace that we see, then, is an eolian constructional surface and not a true alluvial constructional surface. For example, the loess-mantled terrace along the Nishnabotna River at Hancock in Pottawattamie County stands 15 to 20 feet above the present alluvial floodplain. The loess on the terrace is 30 feet thick. Thus, the true alluvial surface of the terrace is 10 to 15 feet below the level of the present alluvial floodplain. If one is interested in the alluvial history of the valley, recognition of the loess surface as a terrace level obviously is incorrect, and one can readily foresee the difficulties that will arise in problems in terrace condition.

The Boyer River Valley is a typical case in point (Fig. 4.16). The present alluvial floodplain ascends from below Lo-

FIG. 4.16. Longitudinal profiles of the present alluvial floodplain and the true alluvial surface of the loess-mantled terrace in the Boyer River Valley in southwest Iowa. (Modified from Daniels and Jordan, 1966.)

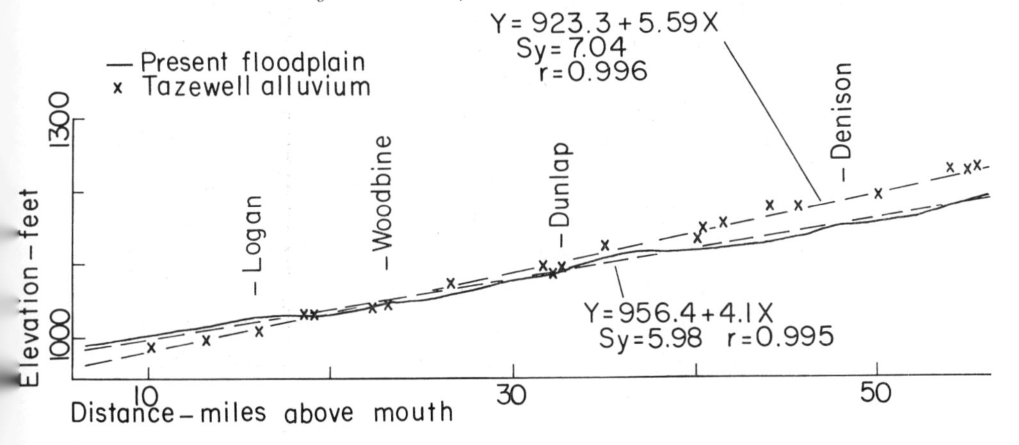

gan in Harrison County to above Denison in Crawford County so systematically that its profile may be expressed as:

$$Y = 956.4 + 4.09X$$

where Y is elevation in feet above sea level and X is distance in miles above the junction of the Boyer and Missouri rivers. The true alluvial surface beneath the loess in the loess-mantled terrace ascends as:

$$Y = 923.3 + 5.59X$$

This means that the gradient of earlier Tazewell alluvial fill was steeper, being 5.59 feet per mile in contrast to the gradient of the present alluvial fill whose gradient is 4.09 feet per mile. The difference of 1.5 feet per mile does not appear to be very much, but it is enough so that the profiles cross near Woodbine in Harrison County (Fig. 4.16). Upstream, Tazewell alluvium and loess stand above the present floodplain. But, downstream (1) the Tazewell alluvium is below the present floodplain; (2) the terrace form preserved by loess only stands above the floodplain; and (3) the present floodplain alluvium laps upward on the loess of the older terrace. These facts show that in Tazewell time, the lower course of the Boyer Valley was not filled as high as at present, and neither was the Missouri Valley to which the Boyer descends. Our analysis shows, by the "*a*" constants of equations, that 33 feet of additional filling took place near the mouth of the Boyer since Tazewell time ($956.4 - 923.3 = 33.1$).

What are soil relations to terraces? In the loess country (pl. 1), soils on loess-mantled terraces generally are very similar to soils formed on loess on the uplands (Corliss and Ruhe, 1955; Fig. 4.4). Of course, soils on loess-free terraces differ from upland soils. The differences are caused mainly by parent material variations in alluvium versus loess, by different topographic situations on the terrace versus the upland, and by differences in age in that the terraces may be considerably younger than the loess upland summits. In the drift province (pl. 1), soil differences may be brought about by the same factors on the terrace versus the upland. We shall look at this problem in a bit more detail later.

Floodplains

As a universal rule, the floodplains of the streams of Iowa are very young regardless of the level of order of the drainageway (Fig. 3.13). In Harrison County in southwest Iowa the Turton channel fill along Thompson Creek is less than 250 years old (W-701; see Table 4.3). The floodplain surface of the Mullenix fill in Thompson Creek and Willow River is less than 1,100 years old (W-799), and the floodplain surface of the Hatcher fill in both valleys is less than 2,020 years old (W-702). Within the Willow River drainage basin, all alluvial fills can be fitted within this time framework. Confluence of the Willow system with the Boyer and Missouri River systems suggests that these systems also belong in the same age brackets.

In southern Iowa the upper floodplain alluvium of the Chariton River is less than 1,800 years old (I-2344). In northeast Iowa, the alluvium of Wolf Creek floodplain is younger than 2,000 years (I-1421). Undoubtedly, as more radiocarbon horizons are analyzed from alluvial deposits in other parts of Iowa, the comparative youth of the floodplains will become known.

Fortunately, these young deposits are leading toward a better understanding of the rapidity of soil formation. At present, quantitative data are scarce on this subject. But, a soil cannot be any older than the surface on which it forms, or older than the material in which it forms. So, if the soils are studied where radiocarbon dates are known, we can begin to quantify the system.

Such a model is on a higher floodplain level along Wolf Creek in Tama County. There, an American elm log was buried in alluvium 9 feet below the surface. As we look at the soils on this surface, we shall see that other things enter the system in addition to the "when" of it. The floodplain has topographic highs and lows, or swells and swales.

Soils on the swells are not as well developed as those in the swales and do not have clayey B horizons with subangular blocky structure. Consequently, the soil horizons are not as distinct on the swells.

FIG. 4.17. Properties of soils on topographic highs (swell) and lows (swale) on floodplain less than 2,000 years old along Wolf Creek in Tama County, Iowa.

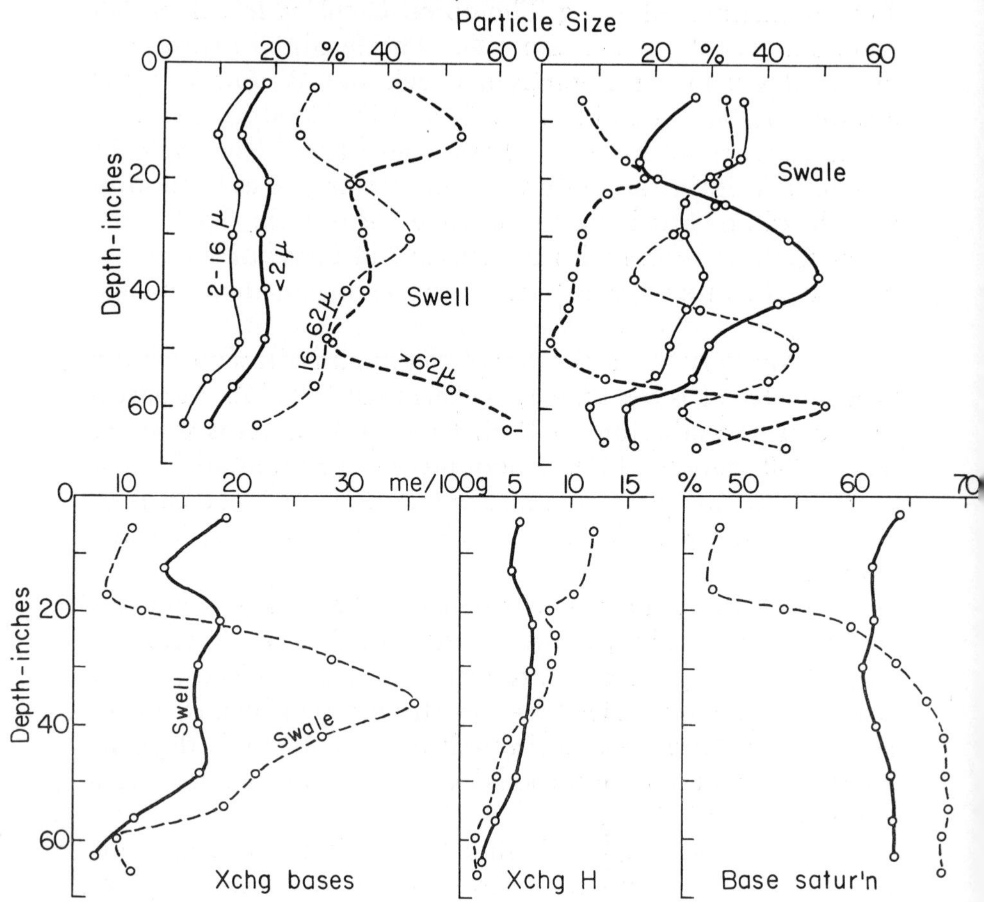

Laboratory data support these conclusions but also introduce many complications in the system. In the soil on the swell, clay distribution is almost straight line to a depth of 50 inches (Fig. 4.17). In the swale soil, clay abundantly increases between depths of 20 and 50 inches, suggesting a textural B horizon. However, sand and coarse silt decrease considerably in the same zone, showing that the clay may be in part the result of stratification during deposition of the alluvium in which the soil formed.

Examination of distributions of exchangeable bases (calcium, magnesium, and potassium), exchangeable hydrogen, and base saturation indicates leaching in the upper 20 inches of the swale soil. Bases are depleted, hydrogen is higher, and base saturation is lower than in subjacent soil horizons. Little variation is noted in the swell soil.

Is any of the clay in the B horizon of the swale soil pedogenic? We can resolve this problem by studying thin sections of samples of soil horizons. Thin sections are made by impregnation of a soil clod with plastic which is permitted to harden. A slice is cut with a diamond saw, and the slice is ground and polished on a lap. The section is studied and may be photographed with the aid of a microscope (Fig. 4.18).

In thin section the swell soil has very little organization of soil components. Sand and silt grains are randomly dispersed in a fine-grained matrix that is impregnated with organic carbon and iron oxide. There is no alignment or orientation of clay. In the swale soil, clay with organic carbon and iron oxide is aligned and oriented around soil voids and around soil aggregates. These coatings are known as "clay skins" which indicate that eluviation and illuviation have occurred as a pedogenic process (Chapter 1). So, now we may conclude that the clay in the zone from 20- to 50-inch depth is in part alluvial depositional and in part illuvial pedogenic. Neither organizational process was active in the less well-developed swell soil.

Another factor is involved in the soil system. The swale is more poorly drained than the swell, so the former environ-

FIG. 4.18. Photomicrographs of soil clods of swell (A–C) and swale (D–F) soils on floodplain of Wolf Creek, Tama County. Scales and depths given. Left profile shows sand and silt grains (light) in unoriented matrix of clay with iron oxide (black) and organic matter (gray). Right profile shows oriented clay, iron oxide, and organic matter, designated by arrows, around voids in B1 (D), B2 (E), and B3 (F) horizons. Note comparable depths of sample horizons.

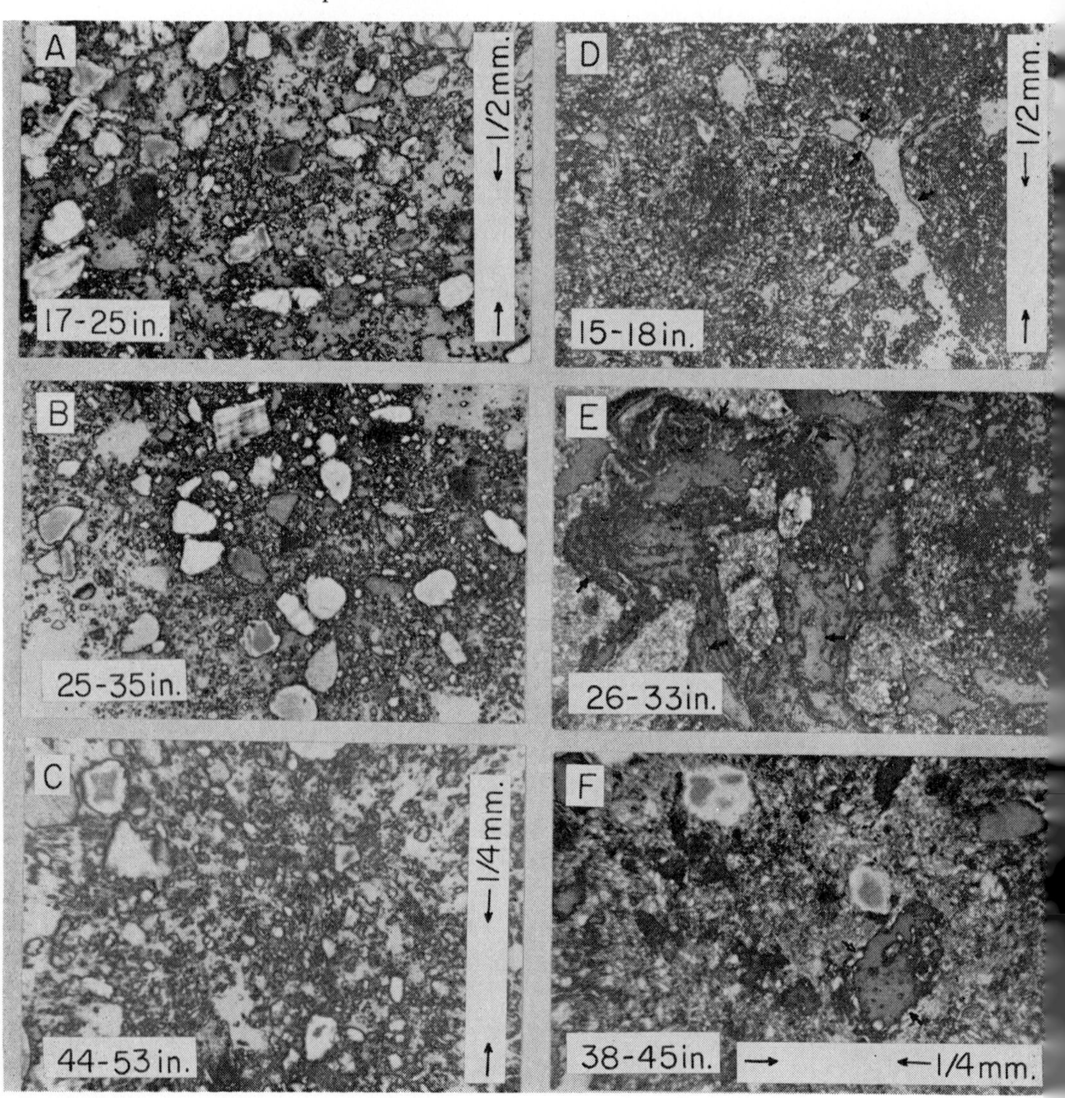

ment is more moist. Soil development increases in the same direction; hence, there must be a relation between intensity of pedogenic process and increase in internal moisture.

How does this all relate to time? On this floodplain of the same age of less than 2,080 years, soils vary from weakly (swell) to more strongly (swale) developed, and the pedogenic system is controlled by inherited differences in parent material and internal drainage of the landscape.

This brief discussion of soils and landscape age is a good departure point for a subject regarding soil chronology and the age of landscape.

SOIL CHRONOLOGY AND LANDSCAPE AGE

We will recall two premises from our criteria of relative dating. (1) A soil is younger than the material in which it forms, and more pertinently, (2) a soil is younger than the surface on which it forms. The second point is the more important of the two and frequently is overlooked. A Nebraskan till may have been exposed on a hillslope 2,000 years ago. Soil formation does not date from the age of the material, that is, Nebraskan time. It dates from the age of the hillslope surface, that is, 2,000 years ago.

A considerable discrepancy may exist between the time as expressed by a chronometer within the soil itself and the age of material and the age of a surface. Organic carbon forms within a soil and generally in the A1 horizon. Commonly, the carbon is translocated downward and accumulates within other horizons in the profile (Chapter 1). The radiocarbon ages of the organic carbon of any soil horizon are readily determined (Table 4.6).

The radiocarbon dates of organic carbon from the A11 horizons of Clarion and Webster soils are 440 ± 120 (L-256A) and 270 ± 120 years (L-256B), respectively. The Clarion is on hillslopes around closed depressions on the Des Moines lobe,

TABLE 4.6. RADIOCARBON DATES OF SOILS IN IOWA

Sample*	Date in years before present	Location	Notes
L-251B	210 ± 130	Hayden Prairie,† Howard County	Organic carbon (OC) from A11 horizon of Cresco-Kenyon soil
L-251C	< 100	Hayden Prairie,† Howard County	Organic carbon from A12 horizon of same soil
L-251D	410 ± 100	Harvard, Wayne County	OC from A1 horizon of Edina soil
L-251E	840 ± 200	Harvard, Wayne County	OC from A2 horizon of same soil
I-(?)	1545 ± 110	Harvard, Wayne County	OC from B2 horizon of same soil
L-256A	440 ± 120	Kalsow Prairie,† Pocahontas County	OC from A11 horizon of Clarion soil
L-256B	270 ± 120	Kalsow Prairie,† Pocahontas County	OC from A11 horizon of Webster soil

* Sample numbers are L for Lamont Geological Laboratory, Columbia University; and I for Isotopes, Inc. Sample number Isotopes value unknown. All dates are strontium carbonate conversions.

† Virgin prairie.

and the Webster is on foot- and toeslopes in the closed depressions (Fig. 4.15). In the previous discussion of the closed system, the hillslope was shown to have stabilized about 3,000 years ago. This dates the land surface at that time. The radiocarbon ages of organic carbon in the soils on the surface are much younger. The discrepancies in age are 2,600 to 2,700 years.

Radiocarbon dates of organic carbon from the A11 and A12 horizons of a Cresco-Kenyon soil are 210 ± 130 (L-251B) and less than 100 years (L-251C), respectively. This soil is on a low gradient hillslope in the loamy surficial sediment area of the Iowan erosion surface in northeast Iowa. In the previous discussion on the Iowan erosion surface, the hillslopes in this region were shown to be as young as 2,900 and 6,100 years. The discrepancies between age of organic carbon of the soil and the surface on which the soil formed can be as much as 2,700 to 5,900 years.

Both of these examples involve virgin soils that are in their natural state. The Clarion and Webster soils are in the Kalsow Prairie in Pocahontas County, and the Cresco-Kenyon soil is in the Hayden Prairie in Howard County. These tracts have been set aside as native prairie reserves.

In a cultivated Edina soil in Wayne County in south central Iowa organic carbon from the A1 horizon is 410 ± 110 years (L-251D), from the A2 horizon is 840 ± 200 years (L-251E), and from the B2 horizon is 1,545 ± 110 years (I-?). The age increases downward in the soil profile. At this site the age of the base of the loess in which the Edina formed is 16,500 years (I-1408; see Table 2.3), and the age of the surface of the loess on which the Edina formed is 14,000 years (Table 2.4). Here, the discrepancies between the dates from the soil and the surface are 12,500 to 13,600 years.

Now, suppose that the Clarion, Webster, Cresco-Kenyon, and Edina soils were buried under loess, and 10,000 years from now the then buried A horizons were sampled and radiocarbon dated. The ages, 10,000 plus 210, 410, or 440 years, would date the time of burial, but they would not date the age of the buried surface. The buried surfaces on which the soils formed could be 13,000, 13,000 to 16,000, and 22,500 to 23,600 years old.

Discrepancies such as these do occur in the radiocarbon chronology in Iowa. In the Bentley section in Pottawattamie County, organic carbon from the A1 horizon of the basal soil in Wisconsin loess is 23,900 ± 1,100 years old (I-1420; see Table 2.3). A piece of spruce wood from the top of the horizon is 21,360 ± 850 years old (I-1023). By subtracting the deviation from the first date and adding the deviation to the second date, the minimum difference in age is about 600 years. Thus, in this case, a spruce tree growing on a soil surface can be considerably younger than a part of the soil and the land surface.

This business of a spruce tree growing on an older land surface in Iowa is a good departure point to our next subject—environment. Spruce, presently, is not native to Iowa. What, then, is the history of vegetation that can be developed within our time framework? We will examine this problem in the next chapter.

CHAPTER 5

ENVIRONMENT

WE CAN EXAMINE THE POSSIBLE changes of environment in the Quaternary in Iowa from several points of view. What can be read from the paleosols? If we study them in comparison to soils on the land surface whose possible environments are better understood, we may infer things about the environments in which the paleosols formed. Certain soil structures are preserved on paleogeomorphic surfaces. Relating these end products to processes active today may also lead to inferences regarding past environment. Weathering zones indicate possible former conditions. The limits of certain glaciations are obvious environmental regimes. The kind of wood that has been used for radiocarbon dating tells us about former vegetation. And last but certainly not least, pollen profiles of the bogs on the land surface permit a reconstruction that approaches the present.

PALEOSOLS

One cannot readily understand or interpret paleosols until he has some knowledge about the conditions under which soils form on the present land surface. The latter is not a simple matter but is a complex interrelationship as shown by a common definition of environment. It is the complex of climatic, edaphic, and biotic factors that act upon an organism or an ecological community and determine its form and survival. So, let us have at the complex and see if we can isolate the factors and determine their influence.

In the present state of the art, it may be possible to read the vegetative environment from the characteristics of a paleosol. It may also be possible to read wetness or dryness. These two factors may permit estimates of past vegetation and climate. But, bear in mind that an interpretation may not be so straightforward. Recall the complex interrelations of factors on the Wolf Creek floodplain. However, we shall try anyway.

Surface Analogues

Soils that form under grass or under trees generally have distinctive properties that are restrictive to either kind of cover (Fig. 5.1). To isolate the effects of vegetation, we will select soils formed in similar Wisconsin loess on the same surface of 14,000 years, whose local topographic sites are similar, and whose present climate is similar. The Tama soil formed under grass, and the Fayette soil formed under forest. The resulting contrasts in profiles of the two soils are pictorially self-evident (Fig. 5.1). The Tama soil has a thick, dark surface horizon that grades downward into a lighter colored B horizon which has pedogenic structure and accumulation of clay. The B horizon grades downward into the C horizon and loess. This is a typical vertical sequence of soil horizons formed under grass.

The Fayette soil has a thin, dark A1 horizon over a gray, ashy-looking, platy A2 horizon which grades downward into a darker colored B horizon which has pedogenic structure and accumulation of clay. The B horizon, in turn, grades downward into the C horizon and loess. The Fayette is known as a Gray-Brown Podzolic soil and the Tama is known as a Prairie soil or a Brunizem.

FIG. 5.1. Profiles of Tama soil formed under grass (left) and Fayette soil formed under forest (right). (Photos by R. W. Simonson.)

TABLE 5.1. COMPARISON OF PROPERTIES OF TAMA AND FAYETTE SOILS OF IOWA

Horizon	Depth	Clay	pH	Organic carbon	Fe_2O_3	Base saturation
	(in.)	*(%)*	*(1:1)*	*(%)*	*(%)*	*(%)*
Tama soil under grass						
Ap	0–6	28.6	5.7	2.35	0.9	66
A1	6–11	32.2	5.8	1.95	1.5	62
A3	11–16	34.2	5.7	1.42	1.6	68
B1	16–20	35.6	5.8	0.97	1.6	72
B21	20–25	35.4	5.7	0.68	1.7	73
B22	25–29	33.2	5.7	0.45	1.7	74
B23	29–35	30.5	5.7	0.34	1.7	76
B3	35–45	28.2	5.8	0.21	1.7	77
C	45–51	28.5	6.1	0.15	1.6	81
C	51–61	27.6	6.5	0.12	1.5	84
Fayette soil under forest						
A1	0–2	13.0	5.6	5.63	0.9	56
A2	2–9	13.5	5.0	0.66	1.0	20
B1	9–17	16.9	5.0	0.30	1.2	34
B21	17–24	23.8	5.0	0.20	1.6	47
B22	24–35	27.7	4.9	0.16	1.7	51
B23	35–42	30.7	5.0	0.17	1.8	64
B3	42–48	29.6	5.2	0.14	1.8	65
C	48 +	26.3	5.1	0.12	1.8	66

If one compares properties from relative positions within the profiles of these soils, distinct differences are noted (Table 5.1). Under forest, eluvial-illuvial processes from A to B horizons are more intense. B to A ratios of clay and iron are 2.27 and 1.8 under forest and 1.04 and 1.13 under grass. Leaching of bases is more intense under forest as shown by an examination of pH and base saturation values which are lower in the forest soil. Organic matter is abundant in only the upper 2 inches of the Fayette. In fact, to a comparable content of 0.66 percent, the depth under forest is only 2 inches, but under prairie is 25 inches.

The form and properties of these soils are so distinctive that one should be able to categorize paleosols as being similar to one or the other.

Wetness or dryness of soils usually are read from the color and patterns of colors in soils (*Soil Survey Manual,* 1951). Again, this is not a straight-across-the-board rule. Red, brown, and yellow colors generally are related to iron oxides, and they usually indicate good internal drainage and aeration within a soil. Nearly pure gray colors or those that even appear blue or green usually indicate poor internal drainage and aeration within a soil in Iowa (Fig. 5.2).

Patterns of color usually are referred to as mottling, which means marked with spots of color. Mottled colors are mainly grays and browns, and usually represent some condition of impeded drainage in the soil, either past or present.

These colors or patterns of color may be within the soil profile or beneath it. A common feature in soils in the thin loess on flat summits in southern Iowa is a zoning of color and color patterns to depth. The solum may be brown or yellowish brown. The next lower zone will be yellowish brown mottled with gray and a downward reversal of gray mottled with brown. The lowest zone is gray with sparse reddish-brown nodules and tubules (Fig. 5.2). The lowest zone is our well-discussed deoxidized zone.

From all of these features we can read, but with caution, the possible vegetative covers and the conditions of wetness or dryness under which the soils formed. Combining these, we may derive the environment of vegetation and climate.

So, let us go back downstairs stratigraphically and examine the paleosols from this point of view. We will restrict ourselves to the paleogeomorphic surfaces just beneath the Wisconsin deposits.

Paleosol Profiles

With the exceptions of great thickness and thorough mineral weathering, little difference is observed between the Yarmouth-Sangamon paleosols and the soils on the present land surface directly above them. As noted previously (Chapter 3), these paleosols are like Planosols and Humic-Gley soils (Ruhe, 1956). Surface soils, such as Edina and Winterset, formed under prairie in conditions of poor internal soil drainage. The paleosols could have formed under similar conditions.

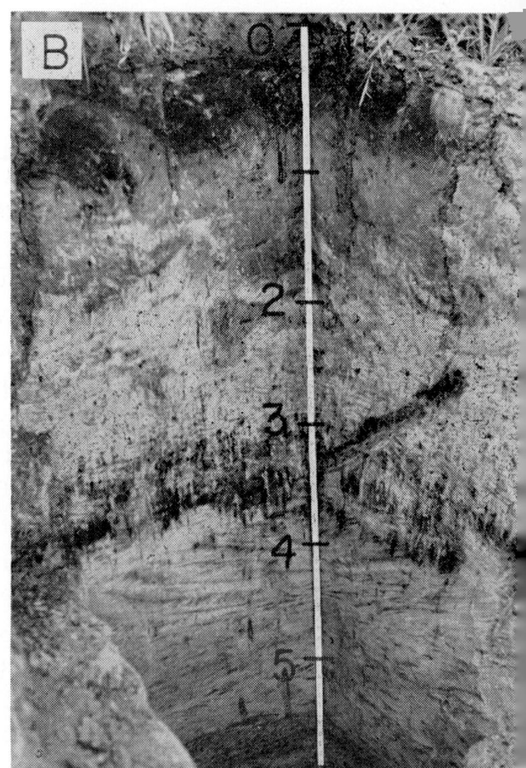

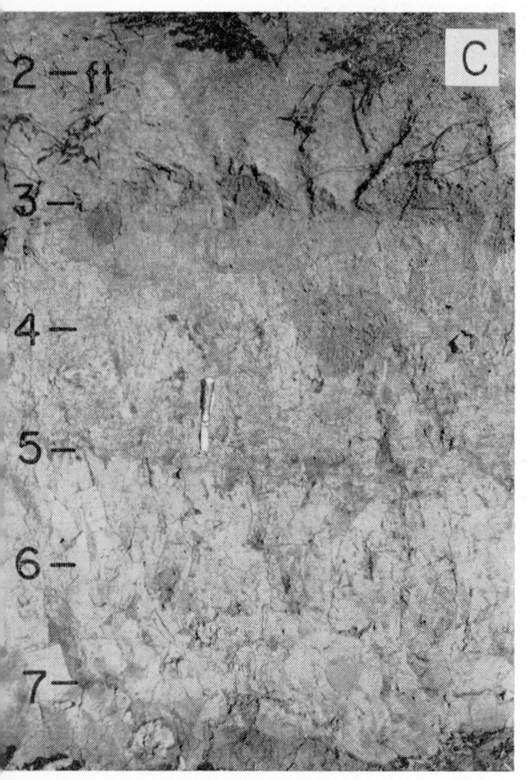

FIG. 5.2 Effect of internal drainage or aeration in the morphology of soils. (A) Well-drained soil with brown and yellowish-brown color below A horizon or depth of 1 foot. (B) Poorly drained soil with light gray to white colors below depth of 2 feet. Reddish-brown (dark-colored) mottles and band of iron oxides at 3 to 4 feet. Light gray to white colors from 4 to 6 feet. (C) Beneath the soil solum which is dark brown and yellowish brown, to depth of 3 feet, is mottled zone of yellowish brown with gray mottles, 3 to 5 feet. Deoxidized zone, 5 to 7 feet, is light gray with reddish-brown tubules. Reddish-brown iron oxide band is at 5 feet. Top of paleosol is at 7 feet.

The poor drainage does not require any climatic change toward greater rainfall in the past, although such change probably occurred. These paleosols are buried on the broad, tabular flats of southern Iowa that exist today. There, the surface soils are the same kinds and are under annual rainfall of 32 to 34 inches. As noted earlier (Chapter 4), this climatic regime is more than adequate to form the surface soils. The main reason for wetness in the soils, past and present, is the topography of the country. The divides are broad and relief is low (Fig. 4.1). Water does not readily drain downward or laterally in this terrain.

The Sangamon paleosols on Loveland loess and Late Sangamon paleosols in Iowa are a different story. They differ distinctly from the surface soils above them which are formed under prairie. They have morphologies akin to soils formed under forest. One need validate this conclusion only by examining the paleosols. Let us turn to Figure 5.1. The soil on the left formed under prairie. The soil on the right formed under forest. Let us turn to the Sangamon paleosol formed in Loveland loess (Fig. 2.1B). In gross morphology, this profile is more like 5.1B than 5.1A. Now let us turn to the Late Sangamon paleosol, Figure 3.4B. This profile formed in Kansan till and a surficial sediment, but even so, its morphology is more like 5.1B than 5.1A.

The Sangamon and Late Sangamon paleosols are analogous to Gray-Brown Podzolic soils that form under forest. They immediately precede the Wisconsin deposits in southwest Iowa. On 58 consecutive ridges between Bentley, Pottawattamie County, and Atlantic, Cass County, these Gray-Brown Podzolic soils are buried in 57 of the ridges (Ruhe and Scholtes, 1956; Simonson, 1954). The surface soils on all ridges are Monona and Marshall soils and associates that formed under prairie. Environment change can be read from the contrasts in these soils. In immediately pre-Wisconsin time the landscape of southwest Iowa was forested. In post-Wisconsin or Recent time, the landscape was in prairie. We shall fill in the gap in time as we proceed, but there are other things about the paleosols that may show the effects of environment.

These paleosols have stronger chromas and redder hues in their B horizons than the surface soils. Here we are referring to soil color that is determined by comparison with the Munsell Soil Color Chart, where all colors are arranged and described by three variables—hue, value, and chroma. Value is the lightness of color, and chroma is its strength or departure from a neutral of the same lightness. A specific color may be 10YR 5/4, where 10YR is the hue, 5 is the value, and 4 is the chroma. The YR indicates that the color is a combination of yellow and red. Most soil colors range from 10R to 5Y hues as 10R, 2.5YR, 5YR, 7.5YR, 10YR, 2.5Y, and 5Y. Yellow increases in this order, but red increases in the reverse order. Pre-Wisconsin paleosols are separable from Wisconsin and Recent surface soils at the hue level. The former generally are 5YR and 7.5YR, whereas the latter usually are 10YR and 2.5Y. The paleosols are redder.

What is the significance of the redder color? If the paleosols are traced southwestward on the Great Plains and beyond the limits of Wisconsin loess, these buried soils emerge on the surface and have colors similar to Reddish-Chestnut and Red-Desert soils (Ruhe, 1965). In this direction the climate becomes warmer and drier. If the buried soils are traced southeastward and down the Mississippi River Valley beyond the limits of Wisconsin loess, the paleosols emerge on the surface and have colors similar to Red-Yellow Podzolic soils. In this direction the climate becomes warmer and more moist. Therefore, a dilemma arises. On the one hand the climate is warmer and drier, and on the other it is warmer and wetter. Common to both regimes is warmer temperature. Does the red color indicate only warmer temperatures during Sangamon and Late Sangamon time? We may get some ideas by looking in different regions.

Along the Rio Grande in southern New Mexico, soils on a surface less than 2,600 years old have color hues of 10YR and 7.5YR (Ruhe, 1967). Soils on surfaces older than 9,550 years have hues of 5YR, 2.5YR, and redder. The dates are radiocarbon values, and herein enters time. On older surfaces, the soils are redder. But, 9,500 or more years ago places the surfaces back in Wisconsin time when glaciation was present in the midcontinent region. Pluvial (wetter) episodes prevailed in the desert. Not only is time involved, but the possibility of more moist conditions arises. Consequently, we must return to our previous statement of "warmer and drier to the southwest of Iowa" and qualify it. It could have been warm and moist, somewhat like the case of "warm and moist to the southeast."

So, we still have time, warmth, and some moisture, and we still have a dilemma. As usual, complications enter any analysis of soil color, and the general conclusion is reached that color alone is not a valid indicator of climate or time alone (Ruhe, 1965). The product of a long interval of time and low climatic intensity could be the same as the product of a short interval of time and high climatic intensity (Simonson, 1954). Within all of these complications, bet on time, warmth, and some moisture for the Sangamon and Late Sangamon episodes in contrast to the following Wisconsin interval in Iowa.

Soil Structures

Certain structural features occur at the base of and within Wisconsin deposits and on surfaces of Wisconsin age. These structures generally are not seen unless an artificial or natural excavation exposes them. They are most distinct when a road cut or road grade is newly made and fresh. All of these structures are known by the general term *patterned ground,* and there are many kinds (Washburn, 1956).

One common structure is arranged in polygonal pattern in plan view (Fig. 5.3C). The polygons are bordered by sand which extends downward in a wedge shape (Fig. 5.3A). In cross section the polygons are broadly domed upward toward an apex, and the wedges extend downward at the contacts between the polygons (Fig. 5.3B).

These structures are very common features on the Iowan

FIG. 5.3. Structures of patterned ground in Iowa. (A) Sand wedge in Nebraskan till (3) below Iowan erosion surface (2) and buried by Wisconsin loess (1). (B) Cross section through polygons that are domed upward between sand wedges (2). Note that stone line (1) follows domed surface. (C) Plan view of sand wedge polygons. (D) Wavy contorted bedding in base of loess above paleosol. (E) Involuted Yarmouth-Sangamon paleosol (2) overlain by Wisconsin loess (1) but also engulfing mass of the loess (3). Photos A to D from Tama County and E from Adair County. All scales in feet.

erosion surface where buried beneath Wisconsin loess in northeast and northwest Iowa and are very common on the Tazewell drift where buried beneath loess in northwest Iowa. These surfaces obviously have cracked up. The sand that fills the cracks was derived during erosion of the till and either washed or drifted into the cracks. The stone line follows the domed and crenulated surface of the till. It does not descend along or into the wedges. Loess covers all and does not fill the wedges. Formation of the polygonal pattern, then, followed cutting of the erosion surface and preceded deposition of the loess. Sand that fills the wedges also covers the erosion surface in the polygons and is beneath the loess. This is the normal position of the sand in this region (Figs. 3.10, 3.11).

Another structure that is also very common is distortion of bedding in the basal part of the loess where a clayey paleosol is slightly lower. The bedding is wavy, crenulated, and sometimes overturned (Fig. 5.3D). The illustration is in a road cut on a flat-summited ridge in Tama County. The Yarmouth-Sangamon paleosol under the loess is also flat lying. Similar features are very common in the railroad cuts from Bentley, Pottawattamie County, to Atlantic, Cass County, where the basal loess also overlies clayey, heavy-textured paleosols.

The contortions of the bedding indicate some kind of flowage of soil in a saturated state which is known by the term *solifluction* (from *solum*, soil, and *fluere*, to flow). This process, as originally defined (Andersson, 1906), is the slow flowing from higher to lower ground of masses of waste saturated with water. As stated, solifluction requires a declivity, higher to lower ground. This is not present in our examples. Their sites are on the level. Perhaps the weight of the overburden of thick loess exerts a downward force that squeezes the saturated loess, causing a lateral flowage and contortion of the bedding.

Another very common structure in Iowa is the abrupt mounding of clayey paleosols into the overlying loess and which may enfold masses of the loess (Fig. 5.3E). The enfolding structures of the paleosol are known as *involutions*. These features are usually restricted to areas where loess is thin. How thin is thin? In Adair County (Fig. 5.3E) the loess is 17 feet thick above the involuted paleosol. In Jefferson County, Iowa, where these structures are abundant (Schafer, 1953), the loess is 7 to 10 feet thick.

What environment can be read from these soil structures? Generally they are attributed to processes active in polar, subpolar, and alpine regions (Washburn, 1956) to frozen ground and intensive frost action (Bryan, 1946), or to the periodic occurrence of a thoroughly effective ground frost (Troll, 1953). But are these restrictions true? They are not necessarily so, and one need no more than be observant in various environmental zones.

It is true that these soil structures occur in cold regions. In Alaska many of these features are along the Denali Road which extends along the south side of the Alaska Range at about 63° N latitude. Here, annual rainfall is about 15 inches and average temperatures in January are about 2° F and in July about 54° F. Minimum temperature can be as low as —54° F and maximum temperature as high as 90° F. About 2 months are frost-free from mid-June to mid-August, and this is the period of heavier rain as 70 percent of the annual amount falls. This is also a region of permafrost, and perennially frozen ground commonly is 1½ to 3 feet below the surface.

Solifluction tongues or lobes are very common features on the hillslopes (Fig. 5.4A). Earth hummocks are abundant (C), and polygons are present throughout the region (E). Cold region processes that give rise to such features (Washburn, 1956) need not be questioned here.

However, very similar structures are observed on hillslopes on Lanai in the Hawaiian Islands. Here, at 21° N latitude the annual rainfall is 15 to 20 inches. The summer is relatively dry, during which about 12 percent of the annual rain falls. The winter is relatively wet, with about 50 percent of the annual precipitation. The temperatures are uniform throughout the year. The January average is 66° F, and the July average is 71° F. Consequently, the whole year is frost-free.

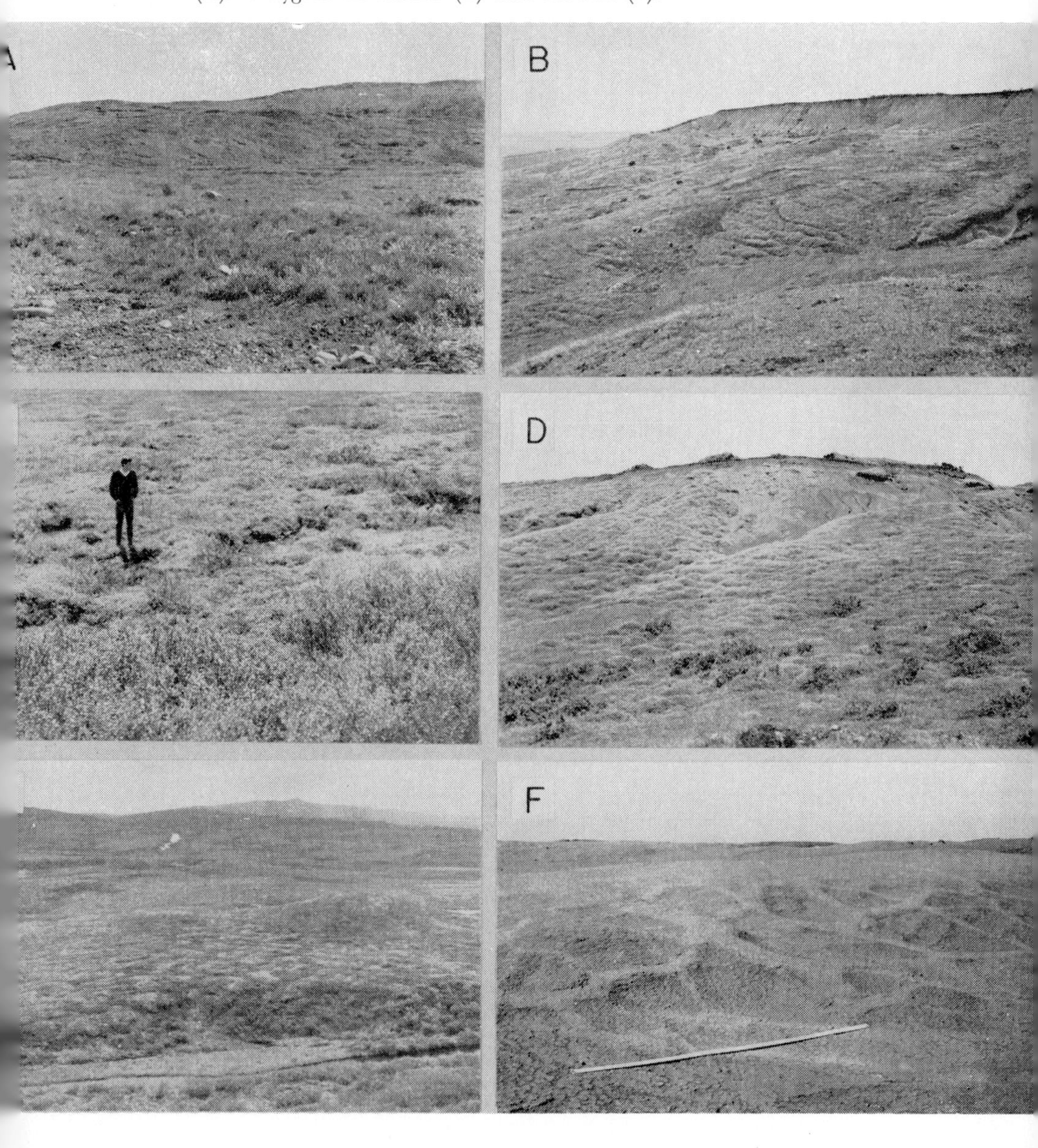

FIG. 5.4. Similar soil structures of patterned ground in different environments. Solifluction tongues or lobes on slopes in Alaska (A) and Hawaii (B). Earth hummocks in Alaska (C) and Hawaii (D). Polygons in Alaska (E) and Hawaii (F).

Solifluction tongues or lobes are very common features on the hillslopes (Fig. 5.4B). Earth hummocks are abundant (D), and polygons are present (F). This is a warm region—not cold—so any association with frozen ground phenomena, past or present, must be ruled out. There may be a relation to wetting and drying. But the similarities of landscape features, cold (Fig. 5.4A to E) to warm (B to F), are very striking indeed. So, we must conclude that either different processes acting under different environments or alike processes independent of environment produce the same results. An interesting tale is involved herein, but study is needed.

Now to get back to the structures of Iowa, how do we interpret the environment? Do we use the cold or warm region analogy or both? Let us fit the structures within our historic framework. The Iowan erosion surface which cracked up into polygons formed in places 22,600 (I-1404) to 18,300 years ago (W-1687; see Table 3.2). Tazewell ice deposited till in northwest Iowa 20,500 (I-1864A) to 20,000 years ago (O-1325; see Chapter 3). As there are 1,780 square miles of that till, there must have been that much ice. In fact, there was probably much more ice because we do not know the limits of the Tazewell drift beneath the younger Des Moines drift lobe. In any event, there was a sizeable refrigerator in northern Iowa during the time that the Iowan erosion surface formed. Therefore, if we are to choose a mechanism for cracking up the Iowan surface into polygons and wedges, cold region processes seemingly are more apropos than warm region processes.

Weathering Zones

We have previously examined the environmental significance of these zones in some detail (Chapter 2). In brief summary, the deoxidized zones are believed to represent cooler and relatively more moist environments that caused a higher stand of ground water than at present. The saturation zones perched on more impermeable paleosols, and did so probably during the time of Cary glaciation and during the postglacial episode of coniferous forest.

Is this history compatible with some of the other soil structures? The contorted bedding in the basal loess (Fig. 5.3D) and the involutions of paleosols engulfing basal loess (Fig. 5.3E) do fit the history. Commonly, in the contorted bedding one observes a thorough mixing of the A horizon of the basal soil of the loess (Table 2.3) and the overlying deoxidized loess. The mixing had to occur after the weathering zones formed. The contorted bedding required saturated flow, and the deoxidized zones required high ground water. They fit together.

In some places the involutions engulf masses of deoxidized loess (Fig. 5.3E). Engulfment had to follow deoxidation of the loess. All of these features could have formed during the colder time of Cary glaciation and can be periglacial features (Schafer, 1953).

The weathering processes forming the deoxidized zones are involved in the problem of the reddish colors of paleosols that underlie the zones (Fig. 2.6A, B). Iron oxide has moved around in the zone and accumulated in nodules and pipestems which are reddish brown. The loess matrix is gray. If iron oxide can move in the zone, it can also move downward out of the zone and accumulate in the lower paleosol. It has done so. For example, a Late Sangamon paleosol beneath a deoxidized zone in Adair County contains hard, yellowish-red and dark reddish-brown iron oxide nodules that are concentrated above the B2 horizon of the paleosol. From the A horizon of the basal soil of the loess downward in the paleosol horizons, the concentrations are: A, 17 percent; paleosol A2, 13 percent; B1, 8 percent; B2, 5 percent; and C, 1 percent. The paleosol B2 horizon acted as a lower, more impermeable barrier above which iron oxide was precipitated.

The questions now are, how much of the red color of the paleosol is truly Late Sangamon, or how much of the red color is due to the iron oxide introduced during burial and Wisconsin age weathering? This is an interesting problem that requires study. In our example, the color of the A2 horizon has not been affected too badly as it is light brownish gray and pale brown of 10YR hue. It seems, then, that environmental interpretation of the paleosol color is not invalidated by the complications of later additions of iron oxide.

GLACIATION

During Wisconsin time in Iowa we know that two glaciations, Tazewell and Cary, have occurred. As an estimate of the area occupied by glacier ice during these times, we can equate the areas of distribution of a given drift. The Tazewell is a bit of a problem because only 1,780 square miles are exposed on the west side of the Des Moines lobe of Cary age (pl. 1). If the Tazewell ice were symmetrically distributed along a north-south axis as is the Cary lobe, and if the Tazewell axis were somewhere near the Cary axis, an estimate could be made of the possible areal coverage of Tazewell ice. This possible area would have occupied the present region of the Tazewell drift plus that part of the Cary lobe bordered by the Altamont moraine and also including that part of the Cary lobe occupied by the Bemis moraine on the west side of the Cary lobe. Thus, 9,670 square miles of northern Iowa could have been occupied by Tazewell ice.

In our previous calculations (Chapter 2), the four phases of the Cary drift lobe, Bemis, Altamont, Humboldt, and Algona, occupy 12,300, 7,030, 3,170, and 1,650 square miles, respectively. The areas of Cary glacier ice would have been similar.

What volumes of ice could have been involved? An average thickness is needed for the third dimension, but is not readily guessed in Iowa. However, the James lobe in South Dakota was a similar glacial feature and was confined in a broad lowland between the higher Prairie Coteau on the east and the Missouri Coteau on the west (Flint, 1955). Thus, low and high elevations affected by Cary ice are measurable, and estimates of ice thickness can be made. The minimum thickness there was 1,000 to 1,300 feet.

Returning to Iowa and assuming an ice thickness of 1,200 feet, the possible volume of Tazewell ice was 2,200 cubic miles; Bemis phase of Cary ice was 2,800 cubic miles; Altamont phase was 1,600 cubic miles; Humboldt phase was 720 cubic miles; and Algona phase was 375 cubic miles. All of these were lots of ice and should have been reasonable refrigerators for ancestral Iowa. The climate should have been somewhat colder. It certainly is around small glaciers of lobate form where they exist today (Fig. 5.5). If one by comparison envisions the maximum Cary ice as a lobe 140 miles long and 110 miles wide (pl. 1), he might feel the cold a little more.

FIG. 5.5. Lobate Worthington glacier near Thompson Pass in Chugach Range near Valdez, Alaska. Terminal moraine extends from right foreground around proglacial lake with icebergs to left middleground. Expand these forms to envision the lobes of Iowa (pl. 1).

How much is "colder"? Again we must resort to a guessing game, but a summary of estimates suggests that mean annual air temperatures were 6° to 8° C (11° to 14° F) colder during glaciation in mid-latitudes than at present (Flint, 1957). Applying these values to present-day Iowa, we need to reduce our mean annual air temperatures as 50° F minus 14° F or 36° F for the southern part of the state, and 46° minus 14° equals 32° F for the northern part of the state. This puts our mean annual temperature down around freezing.

Does this check with regions where glaciers are present today? To find out, we must go to either high latitude or high altitude regardless of latitude, both of which require coldness. We will select high latitude with reasonably low altitude in Alaska south of the Alaska Range. There, glaciers exist from sea level to elevations of a few thousand feet (Fig. 5.5), and, of course, to much higher altitudes. In this region, mean annual air temperatures range from +2° C to −2° C (Péwé, 1965) or 35.6° to 28.4° F. Returning to Iowa, the estimated values of 32° to 36° F fit within the range of temperatures where glaciers exist today.

Within the radiocarbon chronology of Iowa, such colder conditions could have existed from 20,000 to 13,000 years ago. This was the time of Tazewell and Cary glaciation. The history of colder climate can be brought forward toward the present by examining the vegetation record within the radiocarbon framework. Some of this record was previously introduced (Chapter 2), and the details need not be repeated.

PALEOBOTANY

The paleobotanic record within the time framework is based on two kinds of things, the macrofossils and the microfossils of plants. Both kinds are contained in sediments and some of them are associated with paleosols and paleogeomorphic surfaces.

Wood

Wood is preserved in the deposits of Iowa either as trees rooted in place (Fig. 2.4) or as detrital logs. Samples are collected and taken to the laboratory where sections are made and studied under the microscope. Identification of the wood is based on the structural characteristics observed.

It may have been noted in the previous listings of radiocarbon dates (Tables 2.3, 2.4, 2.7, 3.2, 3.3, 4.3) that a considerable number of them are based on the ages of wood samples. When these dates are aligned chronologically, an imposing vegetative history is immediately evident (Table 5.2). The trees are entirely conifers from well back in Pleistocene time of the pre-Wisconsin and through the earlier Wisconsin, the Tazewell and Cary glaciations, and into postglacial and Recent times. On the basis of these dates alone, conifers such as spruce disappear from the historical record at about 11,000 years ago.

The record can be brought closer to the present. In cores extracted from bogs many pieces of spruce and larch wood were penetrated. They were present from the top of the lower muck zone to the base of the lower silt (Chapter 4). Radiocarbon dates at the top of the lower muck zone are as young as 8,000 years (Walker, 1966). As we shall see later, this time was about the beginning of the prairie environment in Iowa.

It appears reasonable that coniferous forest dominated the landscape from the beginning of Wisconsin time, more than 30,000 years ago, until Recent time of about 8,000 years ago. One may wonder what the vegetation looked like. It may have been like the *taiga* of colder regions today (Fig. 5.6). *Taiga* is a Russian word that is applied to the cold, swampy, forested regions of the north. There, spruce dominates the forest stand (Heusser, 1965).

In our previous analysis of present and past Iowa vegetation (Chapter 2), we noted that most of the conifers that are preserved as macrofossils are not native in this state. The nearest similar present-day forests are in northern Minnesota.

TABLE 5.2. RADIOCARBON DATES OF WOOD FROM IOWA

Sample*	Date in years before present	Kind†	Location by county
Recent and postglacial			
W-701	< 250	Box elder	Harrison
W-799	1,100 ± 170	Walnut	Harrison
W-699	1,800 ± 200	Willow	Harrison
I-2334	1,830 ± 100	Red elm	Appanoose
W-702	2,020 ± 200	Red elm	Harrison
I-1421	2,080 ± 115	American elm	Tama
W-700	11,120 ± 440	Spruce	Harrison
W-882	11,600 ± 200	Spruce	Harrison
I-1019	11,635 ± 400	Spruce	Hamilton
I-1862	11,880 ± 170	Larch	Bremer
Glacial: Cary			
C-596‡	11,952 ± 500	Hemlock	Story
C-912	12,120 ± 530	Hemlock	Webster
C-563	12,200 ± 500	Hemlock	Story
I-2333	12,700 ± 290	Spruce	Linn
W-626	12,970 ± 250	Larch	Hancock
C-913	13,300 ± 900	Hemlock	Webster
W-513	13,820 ± 400	Spruce	Greene
I-1268	13,900 ± 400	Spruce	Hamilton
W-517	13,910 ± 400	Spruce	Greene
C-664	14,042 ± 1,000	Hemlock	Story
I-1402	14,200 ± 500	Spruce	Story
W-881	14,300 ± 250	Spruce	Harrison
Glacial: Tazewell			
W-512	14,470 ± 400	Fir, hemlock, larch, spruce	Greene
W-153	14,700 ± 400	Hemlock	Story
I-1270	16,100 ± 1,000	Spruce	Boone
I-1024	16,100 ± 500	Spruce	Polk
C-528	16,367 ± 1,000	Hemlock	Story
W-126	16,720 ± 500	Yew, hemlock, spruce	Polk
C-481	> 17,000	Yew, hemlock, spruce	Polk
W-879	19,050 ± 300	Spruce	Harrison
I-1023	21,360 ± 850	Spruce	Pottawattamie
Wisconsin: pre-Tazewell			
W-141	24,500 ± 800	Larch	Pottawattamie
W-880	37,600 ± 1,500	Spruce	Harrison
Pre-Wisconsin			
W-503	> 29,000	Hemlock	Fayette
W-534	> 34,000	Hemlock	Fayette
W-514	> 35,000	Spruce	Greene
W-516	> 35,000	Spruce	Fayette
W-591	> 37,000	Larch	Cherokee
W-599	> 37,000	Spruce	Linn
W-600	> 37,000	Hemlock	Buchanan
W-139	> 38,000	Hemlock	Buchanan
I-1863	> 39,900	Spruce	Cherokee
I-2758	> 39,900	Larch	Sioux
I-1025	> 40,000	Spruce	Polk

* Sample numbers are C for University of Chicago; I for Isotopes, Inc.; and W for U.S. Geological Survey, Washington, D.C.

† All wood identified by D. W. Bensend, Department of Forestry, Iowa State University.

‡ Chicago dates are by carbon black method.

We also noted previously that a mean annual temperature about 11° cooler than at present in Iowa would approximate the climate of northern Minnesota.

How does this value fit with the calculation of climate needed for Tazewell and Cary glaciations? The latter value was 11° to 14° F cooler than at present. The values fit, and only a slight warming was necessary to change from the Cary glacial environment to the environment of conifers of immediately postglacial and early Recent time.

A considerable gap is present in the record of the macrofossils, and extends from about 8,000 to 2,100 years ago. At this later time, all radiocarbon dates are of wood from kinds of trees that grow in Iowa today (Table 5.2). These kinds are box elder, walnut, willow, red elm, and American elm. Of course this statement may have to be qualified in a few years. The elms may be gone because of the Dutch elm disease.

The gap in the paleobotanic framework can be filled in by pollen stratigraphy of bogs.

FIG. 5.6. Taiga along the Richardson Highway near Gulkana, Alaska. Spruce is prominent.

Pollen

Pollen certainly do not have to be defined, because any short- or long-time resident of Iowa or many other areas knows about the high pollen count during certain seasons of the year and the resultant effects on the well-being of the citizen. What may not be known about pollen is that they accumulate in sediments and are microfossils that may give some clue about vegetation on the landscape during the past and present. Pollen can be identified readily on the basis of observable characteristics of individuals. Study has to be made under a microscope because of size. Pollen of pine and birch may be as small as 0.02 mm in diameter. Maple pollen may be as large as 1.5 mm.

Pollen analysis consists of tabulating the numbers and kinds of pollen in samples collected in vertical series from peat or other sediments. The results generally are plotted in a pollen diagram in which the percentages for each kind of pollen are represented graphically on a horizontal scale at the depth of the sample within the profile. The percentages are presumed to represent indirectly the percentages of kinds of vegetation that grew in the region at the time the sediments were deposited. The basic assumption is that the number of a given kind of pollen deposited per unit of time at a given place is directly related to the abundance of the corresponding kind of vegetation in the surrounding vegetation. Unfortunately, pollen deposition is not so simple, and many complications enter the picture (Davis, 1963). A few obvious complexities are that pollen grains are small and can be transported great distances by wind just as mineral grains of loess have been transported (Chapter 2), or that one kind of plant may produce a large amount of pollen and another plant may produce a small amount in unit time.

However, pollen analysis is widely used to investigate the Quaternary paleobotanic record (Davis, 1965; Cushing, 1965; Whitehead, 1965; Martin and Mehringer, 1965; Heus-

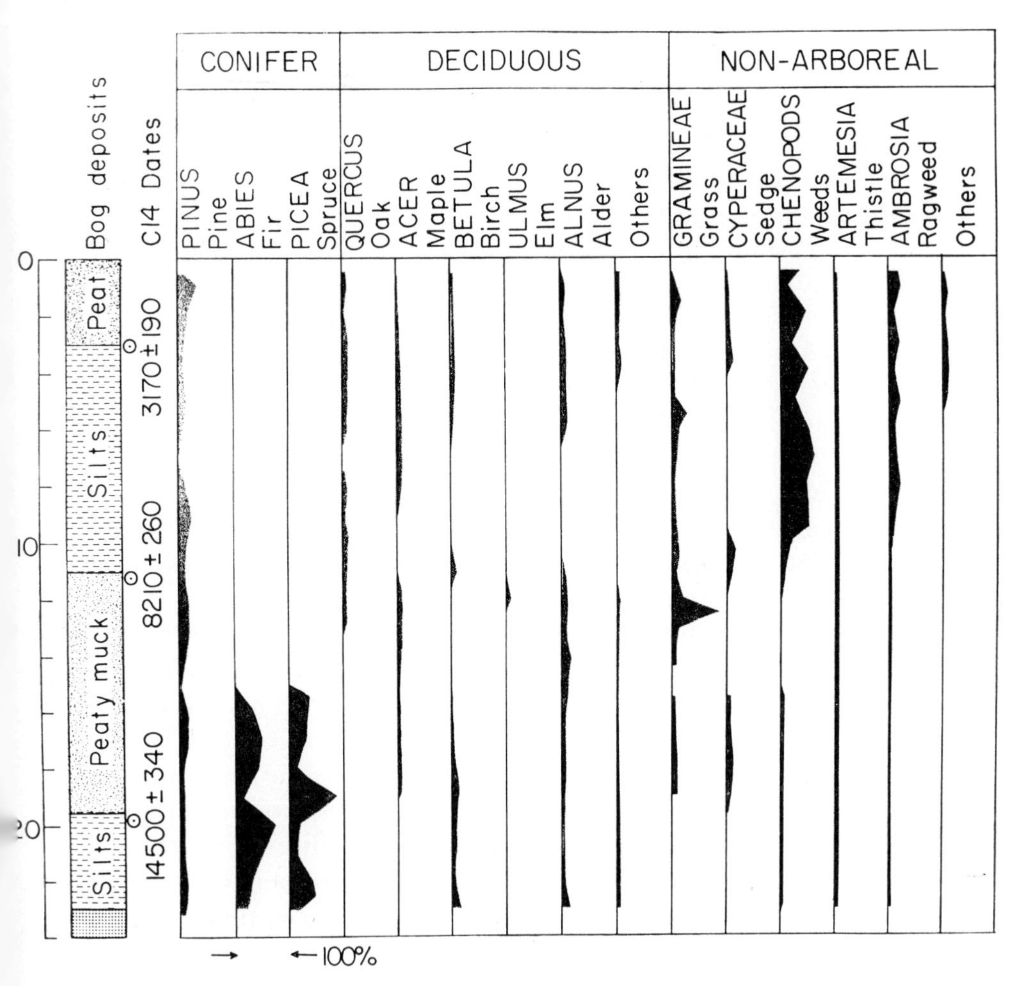

FIG. 5.7. Pollen diagram of McCulloch bog in southern Hancock County, Iowa. (Modified from analysis by Grace S. Brush.)

ser, 1965). Recent comparisons of plant macrofossils with pollen from the same bog show that zoning of the stratigraphic column of the bog could have been done with macrofossils rather than pollen (Watts and Winter, 1966). These studies, done to the north of Iowa in Minnesota, demonstrate reasonable agreement between the larger size and micro plant remains. Consequently, the pollen diagrams give an acceptable record of the past vegetative history.

In Iowa five bogs from the latitude of Ames to near the Minnesota state line have had detailed pollen analysis, and a preliminary note has been published on some of them (Walker and Brush, 1963; Brush, 1968).

Percentage pollen distribution shows remarkable similarity in all bogs, and the McCulloch bog in southern Hancock County can be used as an example (Fig. 5.7). Pollen of conifers such as spruce, fir, and pine are abundant in the lower part of the bog. Usually the spruce and fir are confined to the lower pair of sediments of peaty muck and silts. Commonly, the pollen of deciduous trees such as oak, maple, birch, elm, alder, and others become dominant slightly higher in the bog as the abundance of coniferous pollen wanes. Nonarboreal pollen of grass, sedges, weeds, thistle, ragweed, and others are dominant in the upper silt and upper peat. Microfossils of water lily *(Nymphea)*, cattail *(Typha)*, fern *(Filicineae)*, and mosses *(Lycopsida)* occur throughout the bog. This is to be expected as such aquatic plants live in bogs. The conifers, spruce and fir, usually are lacking in the upper silt and upper peat. However, deciduous trees carry through to the top of the bog sediments.

In a detailed pollen diagram (Fig. 5.7) where depth of bog sediments is from top to bottom and coniferous pollen to the left, deciduous tree pollen in the center, and nonarboreal pollen to the right, the dominance of kind of pollen diagrammatically sweeps from the lower left corner to the upper right corner. Where numerous diagrams need to be examined, the details can become overwhelming. So, summary diagrams can be prepared which show only the percentage sums of coniferous, deciduous, and nonarboreal pollen (Fig. 5.8). The broad

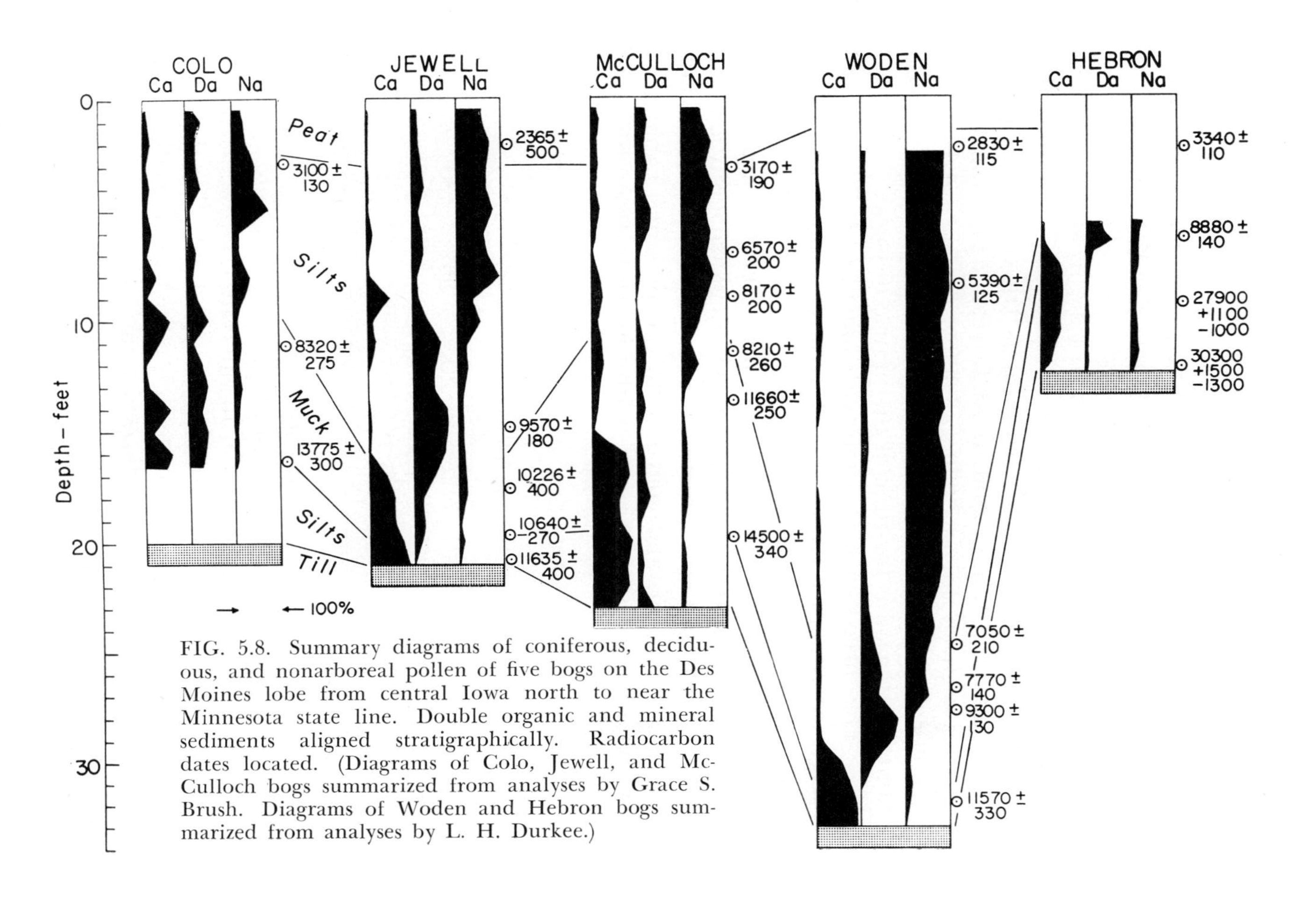

FIG. 5.8. Summary diagrams of coniferous, deciduous, and nonarboreal pollen of five bogs on the Des Moines lobe from central Iowa north to near the Minnesota state line. Double organic and mineral sediments aligned stratigraphically. Radiocarbon dates located. (Diagrams of Colo, Jewell, and McCulloch bogs summarized from analyses by Grace S. Brush. Diagrams of Woden and Hebron bogs summarized from analyses by L. H. Durkee.)

picture is more easily visualized from bog to bog in regard to the pollen distribution in relation to the double organic and mineral stratigraphy of the bogs. One must conclude from the broad picture that the vegetation has changed from coniferous forest to deciduous forest to prairie.

The sequence can be placed in time by insertion of the radiocarbon chronology (Fig. 5.8). Conifers decrease noticeably around 10,000 years ago. Deciduous trees ascend, become dominant, and wane 6,000 to 7,000 years ago. Nonarboreal species begin to ascend 7,000 to 8,000 years ago, are dominant 5,400 and 6,600 years ago, and carry through to the present. This last period is the time of the prairie in Iowa.

The three-fold separation with some overlap may be interpreted climatically as the cool, relatively moist, postglacial episode followed by a warming trend and culminating in the warmer and drier episode of the prairie. Do other features in Iowa fit within this general history? Regional gully cutting and filling began in southwest Iowa 6,080 (M-1071) and 6,800 years ago (W-235) and continue to the present (Table 4.3). These earlier times coincide with the shift from forest to

FIG. 5.9. Summary of vegetational and climatic environments in Iowa inferred from deposits, paleoflora, and other features. Chronology based on radiocarbon dating.

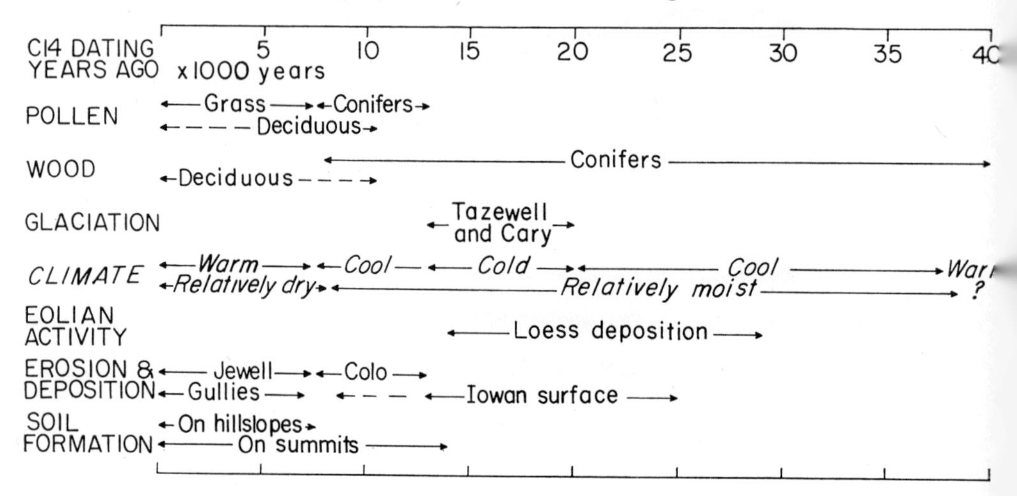

prairie and from relatively more moist to relatively drier climate.

The whole story of environmental change through time can be wrapped up diagrammatically (Fig. 5.9). It should be emphasized that any part of the state of Iowa that was exposed landscape during any part of the time or all of the time would have been subjected to the prevalent environment. For example, during the Bemis phase of Cary glaciation, all of the Des Moines lobe should have been beneath glacier ice, but the rest of Iowa was not covered and should have been in a cold climate under coniferous forest. Accordingly, the question must continually be asked—what was it like here when something was there? The answer will lead to a better understanding of the landscape and the soils on it at a particular place in the state.

PART 2

MEMORANDA

I HAVE CHOSEN THE GENERAL TITLE OF Memoranda for this department because it is plain and simple, and makes no fraudulent promises. I can print under it statistics, hotel arrivals, or anything that comes handy, without violating faith with the reader.

MARK TWAIN
"Introductory to Memoranda"
Life As I Find It

CHAPTER 6

CATALOG OF RADIOCARBON DATES IN IOWA

THE PURPOSE OF THIS CATALOG IS TO record in one place all of the radiocarbon dates in Iowa that have geologic significance and that are currently available at the time of this writing. The dates are listed alphabetically by radiocarbon laboratory and then numerically for a specific laboratory. The format for each date includes the location of the site, a brief narrative description and interpretation of the section, and the names of individuals who collected and submitted the samples. This format follows that of *Radiocarbon,* published annually by the *American Journal of Science* at Yale University. Most radiocarbon laboratories list their dates in this publication, and nine volumes have been published through 1967. The first two volumes were titled *American Journal of Science Radiocarbon Supplement.*

Prior to the establishment of *Radiocarbon* in 1959, most laboratory lists of C14 dates were published in *Science.* Most of the samples from Iowa are scattered through listings in many volumes of *Science* and *Radiocarbon* by numerous laboratories. The following catalog brings together all of the Iowa dates that are geologically meaningful and some of the sites are reinterpreted as required by additional knowledge. Consequently, the listings in the following catalog supersede all prior listings.

There are many other radiocarbon dates from Iowa that are not listed in the catalog. These are mainly archeological in nature, and the lists of many laboratories in *Radiocarbon* describe the sites.

One further point about the catalog is that it may be used as a reference for all the dates that are discussed and tabulated in Part I of this book.

University of Chicago (C)

481 Mitchellville, Polk County *> 17,000*

Spruce wood from loess buried beneath Cary till in east road cut along State Highway 64 in NE¼ sec. 15, T.80N., R.22W. (41°43′N, 93°24′W). Section from surface downward is Cary till, 5 feet; leached Tazewell loess, 3 feet; calcareous loess, 55 feet. Sample from depth of 19.2 feet. Cf. W-126. Collected and submitted by W. H. Scholtes, 1949.

528 Clear Creek, Story County *16,367 ± 1,000*

Spruce wood from loess buried beneath Cary till in south stream cut along Clear Creek in NE¼ SW¼ sec. 5, T.83N., R.24W. (42°1′N, 93°40′W). Section from surface downward is leached Cary till, 2.3 feet; calcareous till, 8 feet; calcareous Tazewell loess with interbedded laminated silts and sands, 31 feet; calcareous till, 2 feet. Sample from depth of 27.3 feet. Cf. W-153. Collected and submitted by W. H. Scholtes, 1949.

596 Cook Quarry, Story County *11,952 ± 500*

Hemlock wood from upper till in east face of Cook Quarry in NW¼ sec. 24, T.84N., R.24W. (42°3′N, 93°35′W). Section from surface downward is leached Cary till, 3 feet; calcareous till, 27 feet; calcareous sand and gravel, 1 foot; calcareous pre-Cary till, 4.7 feet. Sample from depth of 25 feet. Collected by R. E. Wilcox; submitted by R. F. Flint, 1951.

653 Cook Quarry, Story County *12,200 ± 500*

Hemlock wood from same horizon as C-596. Collected by W. Williams and R. Tench; submitted by C. S. Gwynne, 1951.

664 Cook Quarry, Story County *14,042 ± 1,000*

Hemlock wood from sand and gravel between tills. Cf. C-596, C-653.

912 Lizard Creek, Webster County *12,120 ± 530*

Hemlock wood from sands and gravels interbedded between tills in north cut along county road in SE1/4 SW1/4 sec. 10, T.89N., R.29W. (42°32′N, 94°15′W). Section from surface downward is leached sand and gravel, 1.5 feet; calcareous sand and gravel, 8 feet; calcareous Cary till, 7.3 feet; calcareous sand and gravel, 9.5 feet; calcareous till, 3 feet. Sample from depth of 22.5 to 23.5 feet. Collected by R. V. Ruhe and W. H. Scholtes and submitted, 1954.

913 Lizard Creek, Webster County *13,300 ± 900*

Hemlock wood from sands and gravels and same horizon as C-912.

Humble Oil and Refining Company (O)

1325 Cherokee, Cherokee County *20,000 ± 800*

Spruce wood from basal part of Tazewell till in north cut along bypass State Highway 3 in SE1/4 SW1/4 NW1/4 sec. 25, T.92N., R.40W. (43°45′N, 95°34′W). Section from surface downward is oxidized and leached Tazewell till, 7.7 feet; unoxidized and calcareous till, 33.6 feet; zone 27.5 to 41.6 feet contains tree branches and trunks. Description by R. V. Ruhe and W. P. Dietz, 1965. Sample from zone 27.5 to 41.6 feet. Collected by A. R. Dahl and W. D. Frankforter, 1960; submitted by A. R. Dahl, 1961.

Isotopes, Inc. (I)

79 Quimby, Cherokee County *8,430 ± 520*

Charcoal from Quimby site (42°38′N, 95°36′W), 17 feet below surface of bank of Little Sioux River, from hearths in which animals were believed to have been roasted. Extinct *Bison occidentalis* found in association with side-notched points. Collected 1958 by G. A. Agogino and W. D. Frankforter; submitted 1959 by G. A. Agogino, University of Wyoming, Laramie.

295 Effigy Mounds, Allamakee County *1,740 ± 110*

Hickory charcoal from cremation zone of 38- to 40-inch depth in conical mound 12, Effigy Mounds National Monument, sec. 27, T.96N., R.3W. (43°5′N, 91°11′W). Soil profile above zone is weakly developed with A1 horizon, 0–2 inches; A2 horizon, 2–12 inches; B horizon, 12–32 inches. Cremation zone in buried B horizon, 32+ inches. Collected by R. B. Parsons and W. H. Scholtes, 1960; submitted by R. V. Ruhe, 1961.

296 Effigy Mounds, Allamakee County *1,960 ± 90*

Hickory charcoal from zone at 38- to 40-inch depth in conical mound 12. Cf. I-295.

1011 Palermo Area, Grundy County *21,960 ± 1,000*

Organic carbon from silts interbedded between Kansan and Nebraskan tills in center, sec. 29, T.87N., R.17W. (42°19′N, 92°52′W). Core from surface downward was leached Wisconsin loess, 4 feet; calcareous loess, 10 feet; leached Kansan till, 10 feet; calcareous till, 9 feet; leached Aftonian silts with organic carbon, 1 foot; leached Aftonian paleosol, 3 feet. Sample horizon at depth of 33 to 34 feet. Sample converted to strontium carbonate and differs in age from unconverted samples. Cf. I-1265, I-1266, I-1405. Collected by W. P. Dietz and R. C. Shuman, 1961; submitted by R. V. Ruhe, 1963.

1012 Palermo Area, Grundy County *12,350 ± 800*

Organic carbon from basal 4 inches of Wisconsin loess in NE¼ NW¼ sec. 32, T.87N., R.17W. (42°19′N, 92°52′W). Core from surface downward was leached loess, 4 feet; calcareous loess, 10 feet; leached Kansan till, 5+ feet. Sample 0 to 4 inches above till contact. Sample converted to strontium carbonate and differs in age from unconverted samples. Cf. I-1026, I-1404, W-1681. Collected by W. P. Dietz and G. F. Hall, 1962; submitted by R. V. Ruhe, 1963.

1013 Colo Bog, Story County *3,100 ± 130*

Base of upper peaty muck in bog in SW¼ NW¼ sec. 11, T.83N., R.21W. (42°1′N, 93°15′W). Core from surface downward is peaty muck, 3 feet; silts, 7 feet; lower peaty muck, 6 feet; lower silts, 8 feet; Cary till, 3+ feet. Sample at depth of 34 to 36 inches. Grass pollen zone. Collected by P. H. Walker and W. L. Jackson, 1962; submitted by R. V. Ruhe, 1963.

1014 Colo Bog, Story County *8,320 ± 275*

Upper part of lower peaty muck at depth of 11 to 11.25 feet. Hardwood forest pollen zone. Cf. I-1013.

1015 Colo Bog, Story County *13,775 ± 300*

Base of lower peaty muck at depth of 15.5 to 15.75 feet. Conifer pollen zone. Cf. I-1013, I-1014.

1016 Jewell Bog, Hamilton County *2,365 ± 500*

Base of upper peaty muck in bog in NW¼ sec. 19, T.86N., R.24W. (42°14′N, 93°41′W). Core from surface downward is peaty muck, 2 feet; silts, 13 feet; lower peaty muck, 5 feet; lower silts, 9 feet; Cary till, 1+ feet. Sample at depth of 24 to 26 inches. Grass pollen zone. Collected by P. H. Walker and W. L. Jackson, 1962; submitted by R. V. Ruhe, 1963.

1017 Jewell Bog, Hamilton County *10,226 ± 400*

Sample near top of lower peaty muck at depth of 17.5 to 17.75 feet. Hardwood-conifer pollen zone. Cf. I-1016.

1018 Jewell Bog, Hamilton County *10,670 ± 400*

Sample just below base of lower peaty muck at depth of 23.3 to 23.5 feet. Conifer-hardwood pollen zone. Cf. I-1016, I-1017.

1019 Jewell Bog, Hamilton County *11,635 ± 400*

Sample from lower silts just above till contact at a depth of 28 to 28.5 feet. Conifer pollen zone. Cf. I-1016, I-1017, I-1018.

1020 Des Moines, Polk County *17,030 ± 500*

Organic carbon from base of Tazewell loess buried beneath Cary till in road cut at Keosauqua interchange at Interstate Highway 235 in Des Moines (41°35′N, 93°38′W). Section from surface downward is leached Cary till, 2 feet; calcareous till, 6.5 feet; calcareous Tazewell loess with gastropods, 15.3 feet; basal organic zone in loess, 1 foot; Late Sangamon paleosol, 8 feet; leached Kansan till, 5.5 feet. Sample at depth of 24 to 24.5 feet. Sample converted to strontium carbonate and is suspect. Cf. I-1011, I-1012, I-1021, I-1022, I-1026, I-1027. Collected by R. V. Ruhe and W. P. Dietz, 1961; submitted by R. V. Ruhe, 1963.

1021 Casey Paha, Tama County *5,730 ± 250*

Organic carbon from base of Wisconsin loess in east road cut along State Highway 402 in center, sec. 10, T.86N., R.13W (42° 16′N, 92°18′W). Section from surface downward is leached loess, 8 feet; calcareous loess, 10 feet; organic zone in loess, 2 feet; Yarmouth-Sangamon paleosol, 12 feet; leached Kansan till, 11 feet. Sample horizon at depth of 18 to 19 feet. Sample converted to strontium carbonate and differs in age from unconverted sample. Cf. I-1267. Collected by R. C. Shuman, 1962; submitted by R. V. Ruhe, 1963.

1022 Kinross, Keokuk County *20,290 ± 1,000*

Organic carbon and peat from base of Wisconsin loess in SE1/4 SW1/4 sec. 27, T.77N., R.10W. (41°26′N, 91°58′W). Core from surface downward was leached loess, 11 feet 8 inches; Yarmouth-Sangamon paleosol, 1+ feet. Sample from depth of 10.8 to 11.7 feet. Sample converted to strontium carbonate and differs in age from unconverted sample. Cf. I-1406. Collected by E. C. A. Runge, 1962; submitted by R. V. Ruhe, 1963.

1023 Bentley, Pottawattamie County *21,360 ± 850*

Spruce wood from organic zone at base of Wisconsin loess in cut along Rock Island Railroad in NW1/4 sec. 21, T.76N., R.41W. (41°22′N, 95°35′W). Section from surface downward is leached loess, 8 feet; calcareous loess, 29 feet; leached loess, 6 feet; organic zone in base of loess, 3 feet; Sangamon paleosol, 4 feet; leached Loveland (Illinoian) loess, 16 feet. Sample from depth of 43 to 44 feet. Cf. I-1420. Collected by R. V. Ruhe, 1961, and submitted, 1963.

1024 Madrid, Polk County *16,100 ± 500*

Spruce wood from loess interbedded between tills in east road cut of State Highway 60 on south valley slope of Des Moines River in NW1/4 sec. 30, T.81N., R.25W. (41°47′N, 93°49′W). Section from surface downward is leached Cary till, 2 feet; calcareous Cary till, 27 feet; calcareous Tazewell loess with gastropods, 6 feet;

leached Kansan till, 5 feet; calcareous Kansan till with wood fragments in lower foot, 7 feet; calcareous silts with spruce logs in lower 2 feet, 6 feet; calcareous till, 3 feet. Sample from interbedded loess at depth of 30 to 31 feet. Collected by R. V. Ruhe and W. H. Scholtes, 1961; submitted by R. V. Ruhe, 1963.

1025 Madrid, Polk County *> 40,000*

Spruce log from lower silts at depth of 52 feet. Cf. 1024.

1026 Palermo Area, Grundy County *10,370 ± 1,000*

Organic carbon from zone 26–29 inches above base of Wisconsin loess in NE¼ NW¼ sec. 32, T.87N., R.17W. (42°19′N, 92°52′W). Cf. I-1012 for core description. Sample converted to strontium carbonate and differs in age from unconverted samples. Cf. I-1404, W-1681. Collected by W. P. Dietz and G. F. Hall, 1962; submitted by R. V. Ruhe, 1963.

1027 Palermo Area, Grundy County *19,340 ± 900*

Organic carbon from A horizon of paleosol of lower till in the center sec. 29, T.87N., R.17W. (42°19′N, 92°52′W). Cf. I-1011 for core description. Sample at depth from 34.5 to 35 feet. Sample converted to strontium carbonate and differs in age from unconverted samples. Cf. I-1265, I-1266, I-1405. Collected by W. P. Dietz and R. C. Shuman, 1961; submitted by R. V. Ruhe, 1963.

1265 Palermo Area, Grundy County *> 30,000*

Organic carbon from silts between tills in the NW¼ sec. 29, T.87N., R.17W. (42°20′N, 92°51′W). Core from surface downward was leached Wisconsin loess, 8 feet; calcareous loess, 9 feet; calcareous sand, 2 feet; leached Kansan till, 3 feet; calcareous till, 19 feet; leached Aftonian silts, 3 feet; leached Aftonian paleosol 3+ feet. Sample from depth of 42 to 42.5 feet. Sample *not* converted to strontium carbonate. Cf. I-1011. Collected by T. E. Fenton and R. C. Shuman, 1964; submitted by R. V. Ruhe, 1964.

1266 Palermo Area, Grundy County *> 40,000*

Organic carbon from A horizon of paleosol on lower till in the NW¼ sec. 29, T.87N., R.17W. (42°20′N, 92°51′W). Cf. I-1265 for core description. Sample from depth of 44 to 45 feet.

Sample *not* converted to strontium carbonate. Cf. I-1027. Collected by T. E. Fenton and R. C. Shuman, 1964; submitted by R. V. Ruhe, 1964.

1267 Hayward Paha, Tama County *25,000 ± 2,500*

Organic carbon from base of Wisconsin loess above Yarmouth-Sangamon paleosol in center, sec. 36, T.86N., R.13W. (42°13′N, 92°18′W). Core from surface downward was leached Wisconsin loess, 10 feet; calcareous loess, 23 feet; leached loess with organic matter, 3 feet; Yarmouth-Sangamon paleosol, 2+ feet. Sample from depth of 35 to 35.5 feet. Sample *not* converted to strontium carbonate. Cf. I-1021. Collected by T. E. Fenton and R. C. Shuman, 1963; submitted by R. V. Ruhe, 1964.

1268 Stratford, Hamilton County *13,900 ± 400*

Spruce wood from base of Cary till in NW¼ sec. 6, T.86N., R.26W. (42°17′N, 93°56′W). Section in west cut along county road is leached Cary till, 1 foot; calcareous till, 63 feet; interbedded calcareous silts and sands, 7 feet; calcareous pre-Cary till, 21 feet. Sample from depth of 64 to 65 feet. Collected by P. H. Walker, W. L. Jackson, and R. V. Ruhe, 1963; submitted by R. V. Ruhe, 1964.

1269 Salt Creek, Tama County *29,000 ± 3,500*

Organic carbon from base of Wisconsin loess above Yarmouth-Sangamon paleosol in NE¼ NW¼ sec. 7, T.84N., R.14W. (42° 6′N, 92°34′W). Core from surface downward was leached Wisconsin loess, 12 feet; calcareous loess, 30 feet; basal organic zone of loess, 2 feet; Yarmouth-Sangamon paleosol, 12 feet; leached Kansan till, 3+ feet. Sample from depth of 43 to 44 feet. Collected by G. F. Hall and T. E. Fenton, 1963; submitted by R. V. Ruhe, 1964.

1270 Boone, Boone County *16,100 ± 1,000*

Spruce wood from loess interbedded between tills in north cut along new U.S. Highway 30 on west valley slope of Des Moines River in center, sec. 2, T.83N., R.27W. (42°2′N, 93°57′W). Section from the surface downward is leached Cary till, 3 feet; calcareous till, 20 feet; calcareous Tazewell loess with gastropods and wood, 15 feet; leached pre-Wisconsin paleosol, 3 feet; leached Kansan till, 2 feet; calcareous till, 18 feet. Sample from loess at depth of 25 to 25.5 feet. Collected by R. V. Ruhe, 1963, and submitted, 1964.

1402 Nevada, Story County *14,200 ± 500*

Spruce wood from loess interbedded between tills in south cut at Nevada interchange along new U.S. Highway 30 at NC sec. 18, T.83N., R.22W. (41°59′N, 93°27′W). Section from surface downward is leached Cary till, 3 feet; calcareous till, 10 feet; calcareous Tazewell loess with gastropods and wood, 14 feet; calcareous till with stone line at top, 4+ feet. Sample was at depth of 14 feet. Collected by R. V. Ruhe and submitted, 1964.

1403 Grinnell, Poweshiek County *23,900 + 1,100*

Peat at base of Wisconsin loess from foundation excavation beneath Roberts Theater on Grinnell College campus in sec. 9, T.80N., R.16W. (41°44′N, 92°43′W). Section from the surface downward was Wisconsin loess, 12 feet; peat, 2 feet; Yarmouth-Sangamon paleosol. Sample was at depth of 12 to 12.5 feet. Collected by B. F. Graham, 1960; submitted by R. V. Ruhe, 1964.

1404 Palermo Area, Grundy County *22,600 ± 600*

Organic carbon in base of Wisconsin loess at SW corner, sec. 29, T.87N., R.17W. (42°19′N, 92°52′W). Core from surface downward was leached Wisconsin loess, 7 feet; calcareous loess, 11 feet; sand, 1 foot; leached Kansan till, 11 feet; leached Aftonian silts with organic carbon, 2 feet; leached Aftonian paleosol, 5+ feet. Sample from depth of 17.5 to 18 feet. Sample *not* converted to strontium carbonate. Cf. I-1012, I-1026. Collected by R. V. Ruhe and G. F. Hall, 1964; submitted by R. V. Ruhe, 1965.

1405 Palermo Area, Grundy County *>36,000*

Organic carbon from silts interbedded between tills at SW corner, sec. 29, T.87N., R.17W. (42°19′N, 92°52′W). Cf. I-1404 for core. Sample from depth of 30 to 31 feet. Sample *not* converted to strontium carbonate. Cf. I-1011, I-1027, I-1265, I-1266. Collected by R. V. Ruhe and G. F. Hall, 1964; submitted by R. V. Ruhe, 1965.

1406 Kinross, Keokuk County *24,600 ± 1,100*

Organic carbon from base of Wisconsin loess in SE$\frac{1}{4}$ SW$\frac{1}{4}$ sec. 27, T.77N., R.10W. (41°26′N, 91°58′W). Core from surface downward was leached Wisconsin loess, 11.7 feet; Yarmouth-Sangamon paleosol, 2+ feet. Sample from depth of 10.5 to 11 feet. Sample *not* converted to strontium carbonate. Cf. I-1022. Collected by R. I. Dideriksen and J. A. Kovar, 1964; submitted by R. V. Ruhe, 1965.

1408 Harvard, Wayne County *19,200 ± 900*

Organic carbon from base of Wisconsin loess in NW¼ sec. 15, T.68N., R.21W. (40°42′N, 93°16′W). Core from surface downward was leached Wisconsin loess, 8.3 feet; Yarmouth-Sangamon paleosol, 2+ feet. Sample from depth of 7.6 to 8.3 feet. Collected by W. P. Dietz and J. D. Highland, 1964; submitted by R. V. Ruhe, 1965.

1409 Hayward Paha, Tama County *20,300 ± 400*

Organic carbon from base of Wisconsin loess above lower part of Yarmouth-Sangamon paleosol beveled by Iowan erosion surface in center, sec. 36, T.86N., R.13W. (42°13′N, 92°18′W). Core from surface downward was leached Wisconsin loess, 7.5 feet; calcareous loess, 3 feet; leached organic zone in loess, 1.5 feet; truncated Yarmouth-Sangamon paleosol, 6 feet; leached Kansan till, 1 foot. Sample from depth of 11 to 11.5 feet. Collected by T. E. Fenton and W. P. Dietz, 1964; submitted by R. V. Ruhe, 1965.

1410 Murray, Clarke County *20,900 ± 1,000*

Organic carbon from base of Wisconsin loess in NW¼ SW¼ sec. 10, T.72N., R.27W. (41°2′N, 93°57′W). Core from surface downward was leached Wisconsin loess, 11 feet; Yarmouth-Sangamon paleosol, 1+ feet. Sample from depth of 10.3 to 11 feet. Collected by W. P. Dietz and J. D. Highland, 1964; submitted by R. V. Ruhe, 1965.

1411 Greenfield, Adair County *18,700 ± 700*

Organic carbon from base of Wisconsin loess at WC sec. 17, T.76N., R.31W. (41°22′N, 94°27′W). Core from surface downward was leached Wisconsin loess, 15 feet; Yarmouth-Sangamon paleosol, 2+ feet. Sample from depth of 14.3 to 15 feet. Collected by R. V. Ruhe and W. P. Dietz, 1964; submitted by R. V. Ruhe, 1965.

1412 McCulloch Bog, Hancock County *3,170 ± 190*

Base of upper peat in bog in SE¼ sec. 32, T.94N., R.24W. (42°55′N, 93°43′W). Core from surface downward was peat, 3 feet; silts, 8 feet; lower peaty muck, 8.5 feet; lower silts, 4 feet; Cary till. Sample from depth of 36 to 38 inches. Grass pollen zone. Collected by P. H. Walker and W. L. Jackson, 1964; submitted by R. V. Ruhe, 1965.

1413 McCulloch Bog, Hancock County *8,210 ± 260*

Upper part of lower peaty muck. Cf. I-1412. Sample from depth of 11.25 to 11.5 feet. Hardwood forest pollen zone.

1414 McCulloch Bog, Hancock County *14,500 ± 340*

Base of lower peaty muck. Cf. I-1412, I-1413. Sample from depth of 19.3 to 19.5 feet. Conifer pollen zone.

1415 Woden Bog, Hancock County *7,050 ± 210*

Upper part of lower peaty muck in bog at NE corner sec. 13, T.97N., R.26W. (43°13′N, 94°53′W). Core from surface downward was peat, 1.5 feet; silts, 22.5 feet; lower peaty muck, 7.5 feet; lower silts, 1 foot; Cary till. Sample from depth of 21.25 to 21.5 feet. Hardwood forest pollen zone. Collected by P. H. Walker, L. H. Durkee, and W. L. Jackson, 1964; submitted by R. V. Ruhe, 1965.

1416 Woden Bog, Hancock County *11,570 ± 330*

Organic carbon from lower silts. Cf. I-1415. Sample from depth of 31.5 to 32.5 feet. Conifer pollen zone.

1417 Jewell Bog, Hamilton County *9,570 ± 180*

Middle part of lower peaty muck in second core of bog in NW¼ sec. 19, T.86N., R.24W. (42°14′N, 93°41′W). Cf. I-1016. Core from surface downward was peaty muck, 2 feet; silts, 9.5 feet; lower peaty muck, 8 feet; lower silts, 1.5 feet; Cary till. Sample from depth of 14.7 to 15 feet. Conifer-hardwood pollen zone. Collected by P. H. Walker and W. L. Jackson, 1964; submitted by R. V. Ruhe, 1965.

1418 Jewell Bog, Hamilton County *10,640 ± 270*

Base of lower peaty muck. Cf. I-1417. Sample from depth of 19.67 to 20 feet. Conifer pollen zone.

1419A Humeston, Wayne County *16,500 ± 500*

Organic carbon from base of Wisconsin loess in SE¼ sec. 21, T.69N., R.23W. (40°45′N, 93°31′W). Core from surface downward was leached Wisconsin loess, 9.25 feet. Yarmouth-Sangamon paleosol. Sample from depth of 8.5 to 9.25 feet. Carbon residue date. Cf. I-1419B. Collected by W. P. Dietz, J. D. Highland, T. E. Fenton, and R. V. Ruhe, 1964; submitted by R. V. Ruhe, 1965.

1419B Humeston, Wayne County *19,000 + 6,000*
— 3,000

Organic carbon from base of Wisconsin loess. Cf. I-1419A. Humic-acid fraction date. Indicates little natural modern contamination of carbon residue and validity of date of 16,500 years.

1420 Bentley, Pottawattamie County *23,900 ± 1,100*

Organic carbon from buried soil A horizon at base of Wisconsin loess that contained spruce wood. Cf. I-1023 for section description. Reasonable agreement between soil organic carbon and wood dates indicates validity of buried soil organic carbon as dating medium. Collected by R. V. Ruhe, 1964, and submitted, 1965.

1421 Wolf Creek, Tama County *2,080 ± 115*

American elm log buried 9 feet below ground level on north stream bank of Wolf Creek in NE¼, sec. 24, T.86N., R.13W. (42°14′N, 92°18′W). Overlying alluvium noncalcareous. Collected by R. V. Ruhe, W. P. Dietz, and G. F. Hall, 1964; submitted by R. V. Ruhe, 1965.

1852 Woden Bog, Hancock County *2,830 ± 115*

Organic carbon below base of upper peat. Cf. I-1415. Sample from depth of 25 to 28 inches. Grass pollen zone. Collected by P. H. Walker and W. L. Jackson, 1965; submitted by R. V. Ruhe, 1965.

1853 Woden Bog, Hancock County *5,390 ± 125*

Organic carbon from upper silts. Cf. I-1415. Sample from depth of 8 to 8.5 feet. Grass pollen zone.

1854 Woden Bog, Hancock County *7,770 ± 140*

Upper part of lower peaty muck. Cf. I-1415. Sample from depth of 24.25 to 24.5 feet. Hardwood forest pollen zone.

1855 Woden Bog, Hancock County *9,300 ± 130*

Organic carbon from lower silts. Cf. I-1415. Sample from depth of 27.5 to 27.75 feet. Hardwood forest pollen zone.

1856 Hebron Bog, Kossuth County *3,340 ± 110*

Organic carbon from upper part of upper silts in bog in SW¼ NW¼ sec. 27, T.100N., R.27W. (43°27′N, 94°1′W). Core from surface downward was peaty muck, 2 feet; silts, 4.5 feet; lower peaty muck, 1 foot; lower silts, 5 feet; Cary till. Sample from depth of 25 to 28 inches. Collected by P. H. Walker and L. H. Durkee, 1965; submitted by R. V. Ruhe, 1965.

1857 Hebron Bog, Kossuth County *8,880 ± 140*

Organic carbon from silts just above lower peaty muck. Cf. I-1856. Sample from depth of 6.25 to 6.5 feet.

1858 Hebron Bog, Kossuth County *27,990 + 1,100*
— 1,000

Organic carbon from upper part of lower silts. Cf. I-1856. Sample from depth of 9.25 to 9.5 feet. Age is anomalous for lower organic-mineral phase of bogs on Des Moines lobe.

1859 Hebron Bog, Kossuth County *30,300 + 1,500*
— 1,300

Organic carbon from base of lower silts. Cf. I-1856. Sample from depth of 12.25 to 12.5 feet. Age is anomalous for lower organic-mineral phase of bogs on Des Moines lobe.

1860 Sumner Bog, Bremer County *2,930 ± 110*

Peaty muck near base of surface peat in bog in NW¼ NW¼ sec. 13, T.93N., R.11W. (42°52′N, 92°5′W). Core from surface downward is peat, 2 feet; interbedded peaty mucks and silty mucks, 3.25 feet; lower peaty muck, 1.75 feet; lower silty muck, 0.5 foot; alluvium. Sample from depth of 2.25 to 2.5 feet. Grass-deciduous tree pollen zone. Collected by P. H. Walker and W. L. Jackson, 1965; submitted by R. V. Ruhe, 1965.

1861 Sumner Bog, Bremer County *6,130 ± 120*

Peaty muck above lower muck in bog. Cf. I-1860. Sample from depth of 4.5 to 4.75 feet. Tree-grass transition pollen zone.

1862 Sumner Bog, Bremer County *11,880 ± 170*

Larch wood from lower silty muck. Cf. I-1860. Sample from depth of 7.25 to 7.5 feet. Conifer pollen zone.

1863 Mill Creek, Cherokee County *> 39,900*

Spruce wood from lower till in east stream cut of Mill Creek in the center, sec. 32, T.93N., R.40W. (42°49′N, 95°35′W). Section from surface downward is sandy loess, 4 feet; calcareous Tazewell till, 35 feet; interbedded sand, gravel, and silts, 10 feet; calcareous pre-Wisconsin till, 23 feet. Sample from depth of 49 feet. Cf. W-591. Collected by W. D. Frankforter, 1954; submitted by R. V. Ruhe, 1965.

1864A Sheldon, O'Brien County *20,500 ± 400*

Organic carbon from buried soil A horizon beneath Tazewell till in road cut in NE¼ NW¼ sec. 9, T.96N., R.42W. (43°9′N, 95°48′W). Composite section of many sites is loess, 27 inches; calcareous till, 32 inches; calcareous loess, 14 inches; buried soil A horizon, 3 inches; buried soil B horizon to depth of 15+ inches. Possibility of contamination of buried A horizon by roots and translocated organic carbon. Carbon residue value. Cf. 1864B. Collected by R. V. Ruhe and W. P. Dietz, 1965; submitted by R. V. Ruhe, 1965.

1864B Sheldon, O'Brien County *13,400 ± 600*

Humic-acid fraction of 1864A indicating that buried A horizon has been contaminated by younger carbon. However, residue carbon yielded 10 cc CO_2/gm of sample. Humic-acid fraction yielded 0.37 cc CO_2/gm of sample. Calculated δC^{14} of composite of residue and humic acid, i.e., the original sample, is 918 or an age of 20,100 years. This value compared with residue value of 20,500 years (I-1864A) indicates little modern contamination and validity of dates. Collected by R. V. Ruhe and W. P. Dietz, 1965; submitted by R. V. Ruhe, 1965.

1865 Wapello, Louisa County *23,750 ± 600*

Peat from north road cut in EC sec. 5, T.74N., R.3W. (41°14′N, 91°13′W). Section from surface downward is leached Wisconsin loess, 8 feet; leached sands and silts, 4 feet; calcareous sands and silts, 4 feet; peat, 2 feet; leached paleosol B horizon, 1 foot. Sands and silts presumably equivalent to Lake Calvin high terrace sediments. Sample from depth of 17 to 17.5 feet. Cf. OWU-167. Collected by R. V. Ruhe and P. H. Walker, 1965; submitted by R. V. Ruhe, 1965.

2329 4-Mile Creek, Tama County *18,400 ± 310*

Organic carbon and wood fragments from alluvium in tributary valley of 4-Mile Creek in SW¼ sec. 35, T.86N., R.15W. (42°13′N, 92°34′W). Core from surface downward was silty alluvium with sand partings, 14 feet; silty alluvium with organic carbon and wood fragments, 1 foot; bedded silty alluvium with sand partings, 9 feet. Sample depth from 14 to 15 feet. Collected by W. P. Dietz and R. V. Ruhe, 1966; submitted by R. V. Ruhe, 1966.

2330 Elma, Howard County *34,900 + 2,100*
— 1,700

Organic carbon from buried soil A horizon in SE¼ SW¼ sec. 7, T.97N., R.13W. (43°14′N, 92°24′W). Core from surface downward was loam sediment, 3.7 feet; silts, 4.3 feet; buried A horizon, 2.8 feet; buried B horizon, 0.5 foot; leached sandy clay, 3.6 feet; calcareous sandy clay, 1 foot. Sample from depth of 8.7 to 9.3 feet. Collected by W. P. Dietz, 1966; submitted by R. V. Ruhe, 1966.

2331 Davis Corners, Howard County *> 39,900*

Organic carbon from silts interbedded between tills in NW¼ NW¼ sec. 33, T.99N., R.12W. (43°21′N, 92°16′W). Core from surface downward was loam surficial sediment, 2 feet; leached Kansan till, 5 feet; calcareous till, 3 feet; leached Aftonian silts, 3 feet, sandy in upper foot; Aftonian paleosol, 3 feet; leached Nebraskan till, 2.5 feet; calcareous till, 2.5 feet. Sample from depth of 11 to 11.8 feet. Cf. I-1265, I-1266, I-1405. Collected by W. P. Dietz, 1966; submitted by R. V. Ruhe, 1966.

2332 Alburnett Paha, Linn County *20,700 ± 500*

Organic carbon from base of Wisconsin loess in SE¼ NW¼ sec. 14, T.85N., R.7W. (42°10′N, 91°38′W). Core from surface downward was leached Wisconsin loess, 6 feet; calcareous loess, 38 feet; leached Kansan till, 1.5 feet; calcareous till, 11.5 feet; leached Nebraskan till, 4+ feet. Sample from depth of 42.5 to

43.5 feet. Cf. I-1404, W-1681. Collected by W. P. Dietz, T. E. Fenton, and R. V. Ruhe, 1966; submitted by R. V. Ruhe, 1966.

2333 Alburnett Paha, Linn County *12,700 ± 290*

Spruce wood from peat from base of fan alluvium at north basal slope of paha. Cf. I-2332. Core from surface downward was leached silty fan alluvium, 4.5 feet; calcareous fan alluvium, 1.5 feet; peat, 2 feet; calcareous silts with gastropod shells, 2.2 feet; calcareous sand, 2.3 feet; calcareous Nebraskan till, 10+ feet. Sample from depth of 7 to 8 feet. Collected by W. P. Dietz, T. E. Fenton, and R. V. Ruhe, 1966; submitted by R. V. Ruhe, 1966.

2334 Centerville, Appanoose County *1,830 ± 100*

Red elm log beneath 10 feet of alluvium of Chariton River floodplain in WC sec. 27, T.69N., R.17W. (40°44′N, 92°49′W). Sample collected by T. E. Fenton and J. D. Highland, 1966; submitted by R. V. Ruhe, 1966.

2758 Fairview, Sioux County *> 39,900*

Larch wood from mucky silts buried beneath loess and till in north cut along County Road C in SE¼ sec. 11, T.79N., R.48W. (43°14′N, 96°27′W). Section from surface downward is leached Wisconsin loess, 4.8 feet; calcareous loess, 8.2 feet; stone line on till with sand wedges and polygons; calcareous Kansan till, 45 feet; mucky peat with wood, 1 foot, sample horizon; calcareous silts, 2 feet. Collected by R. V. Ruhe and T. E. Fenton, 1967; submitted by R. V. Ruhe, 1967.

3056 Thom Watershed, Tama County *6,200 ± 125*

Organic carbon from alluvium in side valley of tributary of 4-Mile Creek in SE¼ sec. 28, T.86N., R.15W., (42°13′N, 92° 34′W). Section from surface downward is oxidized and leached silt, 4 feet; deoxidized and leached silt, 4 feet; unoxidized and leached silt with charcoal and wood fragments, 3.4 feet; till with leached matrix but with secondary carbonates, 1.5 feet; calcareous till to depth. Sample from depth of 8 to 8½ feet. Collected by W. J. Vreeken, 1967; submitted by R. V. Ruhe, 1967.

3057 Thom Watershed, Tama County *7,710 ± 130*

Organic carbon from base of alluvium in side valley of tributary of 4-Mile Creek. Cf. I-3056. Sample from depth of 10.25 to 10.9 feet. Cf. I-3056 and 3057 with W-235. Dates mark major valley fill that correlates with beginning of prairie in Iowa.

Lamont Geological Observatory, Columbia University (L)

251B Hayden Prairie, Howard County *210 ± 130*

Organic carbon from A11 horizon of Cresco-Kenyon intergrade soil in virgin prairie tract at SE corner, NE¼ sec. 33, T.100N., R.13W. (43°26′N, 92°22′W). Soil A11 horizon, 0–4 inches, 10YR 2/1 Munsell color-moist, fine granular, friable loam to silt loam. Sample converted to strontium carbonate. Collected by R. V. Ruhe and F. F. Riecken, 1954; submitted by Soil Survey Laboratory, Soil Conservation Service, Beltsville, Maryland, 1955.

251C Hayden Prairie, Howard County *< 100*

Organic carbon from A12 horizon of Cresco-Kenyon intergrade soil. Cf. L-251B. Soil A12 horizon, 4–8 inches, 10YR 2/1, fine granular, friable, loam to silt loam. Sample converted to strontium carbonate.

251D Harvard Site, Wayne County *410 ± 100*

Organic carbon from A1 horizon of Edina silt loam at NW corner, SW¼ SW¼ sec. 10, T.68N., R.21W. (40°40′N, 93°16′W). Soil A1 horizon, 0–6 inches, 10YR 2/1, medium granular silt loam. Sample converted to strontium carbonate. Collected by R. V. Ruhe and F. F. Riecken, 1954; submitted by Soil Survey Laboratory, SCS, Beltsville, Maryland, 1955.

251E Harvard Site, Wayne County *840 ± 200*

Organic carbon from A2 horizon of Edina silt loam. Cf. L-251D. Soil A2 horizon, 8–12 inches, 10YR 3/1, weak medium to coarse platy silt loam. Sample converted to strontium carbonate.

256A Kalsow Prairie, Pocahontas County *440 ± 120*

Organic carbon from A11 horizon of Clarion loam in virgin prairie at SE corner, NE¼ sec. 36, T.90N., R.31W. (42°34′N, 94°27′W). Soil A11 horizon, 0–6 inches, 10YR 2/1, granular loam.

Sample converted to strontium carbonate. Collected by F. F. Riecken and F. J. Carlisle, 1954; submitted by Soil Survey Laboratory, SCS, Beltsville, Maryland, 1955.

256B Kalsow Prairie, Pocahontas County *270 ± 120*

Organic carbon from A11 horizon of Webster silty clay loam. Cf. L-256A. Soil A11 horizon, 0–6 inches, 10YR 2/0, granular silty clay loam. Sample converted to strontium carbonate.

University of Michigan (M)

932 Turin, Monona County *4,720 ± 250*

Skeleton number 3, adolescent flexed on left side, in burial pit sprinkled with red ochre and associated with side-notched projectile point and string of gastropod shell beads. At depth of 13 to 14 feet in gully fill in loess exposed in wall of gravel pit in SE¼ sec. 4, T.83N., R.44W. (42°1′N, 95°57′W). Incorrectly reported as buried in sandy to silty loess. Collected by R. J. Ruppe and W. D. Frankforter, 1955; submitted by R. J. Ruppe, 1956.

984 Pony Creek, Mills County *7,250 ± 400*

Charcoal from hearth 17 feet below surface of alluvial terrace in sec. 2, T.73N., R.43W. (41°9′N, 95°47′W). Collected by D. Henning, 1958; submitted by W. D. Frankforter, 1958.

1071 Turin, Monona County *6,080 ± 300*

Lumbar vertebrae and sacrum from *Bison* sp. in gully fill in loess at depth of 8 feet below human skeleton. Cf. M-932. Collected and submitted by W. D. Frankforter, 1955.

Ohio Wesleyan University (OWU)

167 Wapello, Louisa County *23,050 ± 820*

Woody peat beneath alluvium correlative of Lake Calvin high terrace. Cf. I-1865 for section description and agreement of dates. Collected and submitted by J. G. Ogden, 1965.

U.S. GEOLOGICAL SURVEY, WASHINGTON, D.C. (W)

126 Mitchellville, Polk County *16,720 ± 500*

Spruce wood from Tazewell loess buried beneath Cary till. Cf. C-481 for location and section description. Collected by R. V. Ruhe and W. H. Scholtes, 1956; submitted by R. V. Ruhe, 1957.

139 Independence, Buchanan County *> 38,000*

Hemlock wood from leached silt interbedded between tills in north cut along U.S. Highway 20 in NW1/4 sec. 3, T.88N., R.8W. (42°28'N, 92°46'W). Section from surface downward is leached loam surficial sediments, 3.2 feet; stone line; leached Kansan till, 5.7 feet; calcareous till, 2 feet; leached Aftonian silt, 2 feet; peat, 1 foot; leached silt containing wood, 1.8 feet; calcareous Nebraskan till, 2 feet. Sample from depth of 14 to 15 feet. Collected by R. V. Ruhe and W. H. Scholtes, 1955; submitted by R. V. Ruhe, 1956.

141 Hancock, Pottawattamie County *24,500 ± 800*

Larch wood from peat at base of Wisconsin loess in south cut along Rock Island Railroad in NC NE1/4 sec. 23, T.76N., R.40W. (41°21'N, 95°25'W). Section from surface downward is leached Wisconsin loess, 9.7 feet; calcareous loess, 18.3 feet; leached loess, 1.9 feet; peat, 1.3 feet; leached silt, 1.5 feet; Sangamon paleosol, 7.7 feet; Loveland loess, 5+ feet. Sample from depth of 30 to 31 feet. Collected and submitted by R. V. Ruhe, 1954.

153 Clear Creek, Story County *14,700 ± 400*

Hemlock wood from Tazewell loess buried beneath Cary till. Cf. C-528 for location and section description. Collected by R. V. Ruhe and W. H. Scholtes, 1954; submitted by R. V. Ruhe, 1955.

235 Middle Silver Creek, Pottawattamie County *6,800 ± 300*

Organic carbon from gully fill 17 feet thick in north cut along Rock Island Railroad in SC sec. 13, T.76N., R.41W. (41°21'N, 95°31'W). Gully truncates Wisconsin loess and Sangamon paleosol. Sample from depth of 16 to 17 feet. Sample converted to strontium carbonate. Collected by R. V. Ruhe, 1954; submitted by Soil Survey Laboratory, SCS, Beltsville, Maryland.

503 Fayette, Fayette County *> 29,000*
Hemlock wood from basal part of upper till in east cut along the Milwaukee Railroad in the C NE1/4 sec. 3, T.92N., R.8W. (42°48′N, 92°46′W). Section from surface downward is leached Kansan till, 9.3 feet; calcareous till, 10.3 feet; oxidized and calcareous Nebraskan till, 2.5 feet. Sample from depth of 18 to 19.5 feet. Collected by R. V. Ruhe and W. H. Scholtes, 1955; submitted by R. V. Ruhe, 1956.

512 Scranton No. 1, Greene County *14,470 ± 400*
Spruce wood from Tazewell loess buried beneath Cary till in north cut along U.S. Highway 30 in SW1/4 sec. 33, T.84N., R.31W. (42°2′N, 94°27′W). Section from surface downward is leached Cary till, 3.5 feet; calcareous till, 14.3 feet; calcareous Tazewell loess with gastropods and trees rooted in place, 17.7 feet; calcareous pre-Wisconsin till, 13.5 feet. Sample from depth of 17.8 to 18.8 feet. Collected and submitted by R. V. Ruhe, 1956.

513 Scranton No. 1, Greene County *13,820 ± 400*
Spruce wood from Tazewell loess buried beneath Cary till. Cf. W-512. Sample from depth of 34 to 35 feet.

514 Scranton No. 1, Greene County *> 35,000*
Spruce wood from lower till. Cf. W-512. Sample from depth of 48 to 49 feet.

516 Maynard, Fayette County *> 35,000*
Spruce wood from upper till in east cut along Rock Island Railroad in NE1/4 sec. 27, T.92N., R.9W. (42°45′N, 91°53′W). Section from surface downward is oxidized and leached till, 6.8 feet; oxidized and unleached till, 5 feet; unoxidized and unleached till, 5+ feet. Sample from depth of 14.3 feet. Till was presumably "Iowan" but is Kansan or Nebraskan. Collected by R. V. Ruhe and W. H. Scholtes, 1955; submitted by R. V. Ruhe, 1956.

517 Scranton No. 2, Greene County *13,910 ± 400*
Spruce wood from tree rooted in place in loess interbedded between tills in east cut along county road in NE1/4 sec. 26, T.84N., R.32W. (42°3′N, 94°33′W). Section from surface downward is leached Cary till, 3 feet; calcareous Cary till, 40.9 feet; cal-

careous Tazewell loess with weakly formed buried soil in upper 19 inches, 4 feet; stone line; calcareous Kansan till, 16.1 feet; leached Nebraskan till, 1.9 feet; leached sands and gravels, 5 feet. Sample from depth of 43.9 to 45.4 feet. Collected by R. V. Ruhe and W. H. Scholtes and submitted, 1956.

534 Fayette, Fayette County *> 34,000*
Hemlock wood from presumed "Iowan" till. Rerun of W-503, cf. Till is Kansan or Nebraskan.

548 McCulloch Bog, Hancock County *11,660 ± 250*
Organic muck from base of bog. Cf. I-1412 for location. Boring from surface downward was peat, 4.5 feet; peaty muck, 2.5 feet; muck, 7 feet. Sample from depth of 12.5 to 13 feet. Residue carbon value. Conifer pollen zone. Cf. I-1414. Collected by R. V. Ruhe and W. H. Scholtes and submitted, 1956.

549 McCulloch Bog, Hancock County *8,170 ± 200*
Organic muck from bog. Cf. W-548, I-1413. Sample from depth of 8.5 to 9.5 feet. Residue carbon value. Hardwood-conifer pollen zone.

551 McCulloch Bog, Hancock County *6,570 ± 200*
Peaty muck from bog. Cf. W-548, I-1413. Sample from depth of 6.5 to 7.5 feet. Residue carbon value. Grass pollen zone.

552 McCulloch Bog, Hancock County *11,790 ± 250*
Humic-acid fraction of W-548. NaOH soluble.

553 McCulloch Bog, Hancock County *8,110 ± 200*
Humic-acid fraction of W-549. NaOH soluble.

554 McCulloch Bog, Hancock County *6,580 ± 200*
Humic-acid fraction of W-551. NaOH soluble.

557 McCulloch Bog, Hancock County *10,050 ± 250*
Strontium carbonate conversion of organic matter of W-548. Cf. W-548 and W-552 and note difference in ages. (See Rubin and Alexander, 1960.) Collected by R. V. Ruhe and W. H. Scholtes, 1955; submitted by Soil Survey Laboratory, SCS, Beltsville, Maryland, 1956.

591 Quimby, Cherokee County *> 37,000*

Larch wood from lower part of upper till in north cut along State Highway 31 in SE corner, sec. 5, T.90N., R.40W. (40°38′N, 95°36′W). Section from surface downward is leached Wisconsin loess, 2.5 feet; stone line; calcareous till, 41.4 feet; calcareous silts, 3.5 feet; leached paleosol, 6 feet. Sample from depth of 40.9 to 43.9 feet. Upper till presumed to be "Iowan" and lower till Kansan. Tills are Kansan and Nebraskan. Collected by R. V. Ruhe and W. H. Scholtes and submitted, 1956.

599 Central City, Linn County *> 37,000*

Spruce wood in upper till in east cut of county road in C sec. 21, T.86N., R.6W. (42°14′N, 91°33′W). Section from the surface downward is leached surficial loam sediment, 3.5 feet; leached sand and gravel, 1.7 feet; leached Kansan till, 4.8 feet; calcareous till, 5.1 feet; calcareous sand and gravel, 1+ feet. Sample from depth of 14 to 15 feet. Till was presumed to be "Iowan." Collected by R. V. Ruhe and W. H. Scholtes and submitted, 1957.

600 Independence, Buchanan County *> 37,000*

Hemlock wood from silt interbedded between tills. Cf. W-139 for location and section description. Sample from depth of 11 to 12 feet. Collected by R. V. Ruhe and M. Rubin and submitted, 1957.

625 Britt, Hancock County *13,030 ± 250*

Peat from buried soil in outwash in west bank of drainage ditch in SE¼ sec. 8, T.96N., R.25W. (43°8′N, 93°49′W). Section from surface downward is interbedded calcareous silts and sands, 8.3 feet; leached woody peat, 0.7 foot; leached silt, 0.2 foot; calcareous sand and gravel, 5 feet. Sample from depth of 8.3 to 9 feet. Collected by R. V. Ruhe and W. H. Scholtes and submitted, 1957.

626 Britt, Hancock County *12,970 ± 250*

Larch wood from buried tree stump rooted in place. Cf. W-625. Sample from depth of 5.5 to 6.5 feet.

699 Thompson Creek, Harrison County *1,800 ± 200*

Willow wood from buried tree stump rooted in place in east bank of Fox Branch of Thompson Creek in SW¼ SE¼ sec. 13,

T.80N., R.43W. (41°44'N, 95°47'W). Section from surface downward is leached silty alluvium, 2.5 feet; calcareous silts, 22.3 feet; base of upper Mullenix bed; calcareous silts of lower Hatcher bed, 5.7 feet. Sample from depth of 12 to 14 feet. Collected and submitted by R. B. Daniels, 1957–1958.

700 Thompson Creek, Harrison County *11,120 ± 440*

Spruce wood from near top of Soetmelk bed in south bank of Thompson Creek in SW¼ SW¼ sec. 13, T.80N., R.43W. (41°44'N, 95°47'W). Section from surface downward is calcareous postsettlement alluvium, 3 feet; leached silt of Mullenix bed, 1.5 feet; calcareous silt, 6 feet; calcareous silts of Hatcher bed, 13.8 feet; calcareous silts of Watkins bed, 9.7 feet; calcareous silts of Soetmelk bed with twigs and logs, 1 foot. Sample from depth of 34 to 35 feet. Collected and submitted by R. B. Daniels, 1957–1958.

701 Thompson Creek, Harrison County *< 250*

Box-elder wood from buried tree stump rooted in place in channel fill in east bank of Thompson Creek in NW¼ NE¼ sec. 18, T.80N., R.42W. (41°44'N, 95°46'W). Section from surface downward is leached channel fill silts of Turton bed, 8.2 feet; calcareous channel fill silts, 3 feet; calcareous silts of Mullenix bed, 7.3 feet. Sample from depth of 9.3 feet. Collected and submitted by R. B. Daniels, 1957–1958.

702 Thompson Creek, Harrison County *2,020 ± 200*

Red-elm wood from log in alluvium in east bank of Willow River at junction of Thompson Creek in SW¼ SE¼ sec. 14, T.80N., R.43W. (41°44'N, 95°47'W). Section from surface downward is leached silts of Mullenix bed, 10.5 feet; calcareous silts, 6.1 feet; calcareous silts of Hatcher bed, 14.3 feet. Sample from depth of 20.5 to 21.5 feet. Collected and submitted by R. B. Daniels, 1957–1958.

799 Thompson Creek, Harrison County *1,100 ± 170*

Walnut wood from log in alluvium of Mullenix bed on north bank of Thompson Creek. Cf. W-700 for location and section description. Collected and submitted by R. B. Daniels, 1958–1959.

879 Logan, Harrison County *19,050 ± 300*

Spruce wood from banded alluvium beneath Tazewell loess

in north face of limestone quarry in NW¼ sec. 20, T.79N., R.42W. (41°39′N, 95°45′W). Section from surface downward is calcareous Tazewell loess, 16 feet; leached bedded silts, 6.2 feet; leached buried A horizon, 0.5 foot; leached bedded clayey silts, 6 feet; leached buried A horizon, 0.5 foot; leached interbedded clayey silts and sands, 0.8 foot; leached buried A horizon, 0.5 foot; leached clayey silts, 2.1 feet; leached interbedded clayey silts and sands, 5 feet; limestone bedrock. Sample from depth of 19.4 feet. Collected by R. B. Daniels, J. R. Rubin, and M. Rubin and submitted, 1959.

880 Logan, Harrison County *37,600 ± 1,500*

Spruce wood from lowest buried A horizon in alluvium beneath Tazewell loess. Sample from depth of 30 to 30.5 feet. Cf. W-879.

881 Willow River, Harrison County *14,300 ± 250*

Spruce wood from log buried in alluvium in west bank of Willow River in SW¼ sec. 11, T.80N., R.43W. (41°45′N, 95°47′W). Section from surface downward is leached postsettlement silts, 2.5 feet; leached Mullenix silts, 7 feet; calcareous silts, 6.2 feet; calcareous Hatcher silts, 5.9 feet; calcareous Watkins silts, 11.1 feet; calcareous Soetmelk silts, 1.3 feet. Sample from depth of 32.7 to 33 feet. Collected and submitted by R. B. Daniels, 1959.

882 Willow River, Harrison County *11,600 ± 200*

Spruce wood from Watkins alluvium. Cf. W-881. Sample from depth of 28.4 to 29.4 feet.

1618 Palermo Area, Grundy County *21,600 ± 600*

Organic carbon from base of Wisconsin loess. Duplicate sample of I-1404. Cf. for location and core description.

1687 Salt Creek, Tama County *18,300 ± 500*

Organic carbon from base of Wisconsin loess above Iowan erosion surface on Kansan till in SC sec. 6, T.84N., R.14W. (42°6′N, 92°31′W). Core from surface downward was leached Wisconsin loess, 9 feet; calcareous loess, 6 feet; calcareous sand, 0.5 foot; Kansan till. Sample from depth of 14.2 to 15 feet. Collected by G. F. Hall and E. Robello, 1966; submitted by R. V. Ruhe, 1966.

CHAPTER 7

CONTAMINATION

POSSIBLE CONTAMINATION OF RADIOcarbon samples has been mentioned sporadically throughout the preceding text (Part I) and in the descriptions of radiocarbon sites (Chapter 6). Contamination may be caused naturally in the field by the penetration of younger plant roots into the radiocarbon horizon or by the downward movement of more soluble humified materials broadly classified as humic acids (Olson and Broecker, 1958). Such contamination may be present in peats, mucks, sediments, and in the A horizons of buried soils. In sediments, some of the mineral detritus may be older organo-mineral particles such as carbonaceous shale. These inclusions in a bulk sample of sediment may also cause erroneous dating. Last, preparation of samples in the laboratory may also lead to contamination of the sample.

These problems have been investigated through the years where need has arisen in establishing the radiocarbon chronology of Iowa. The validity of certain dates became questionable simply on the basis of the internal consistency of the whole chronology.

THEORETICAL CONSIDERATIONS

The magnitude of the effects of contamination by younger carbon in an older sample can be determined by mathematically manipulating the exponential radioactive decay equations (Chapter 1). The following discussion follows a study on the sediments of the bogs of the Des Moines lobe in Iowa (Walker, P. H., "Soil and Geomorphic History in Selected Areas of the Cary Till, Iowa," unpublished Ph.D. thesis, Iowa State University, Ames, 1965).

Our first-order equation is (Chapter 1)

$$-\frac{dN}{dt} = k_1 N$$

By integrating

$$-\int_{N_o}^{N} \frac{dN}{N} = k_1 \int_{0}^{t} dt$$

from which

$$\ln\left(\frac{N_o}{N}\right) = k_1 t$$

where N_o is the modern radioactivity in disintegrations per minute per gram of carbon, and N is the radioactivity of the unknown sample of carbon of age t years.

Using the half-life ($t_{1/2}$) of carbon of 5,730 years, then

$$N = 15.3e^{-\frac{0.693t}{5730}}$$

If one gram of carbon contains some old carbon (Ca) and some young carbon (Cm), the number of disintegration counts (Nc)

will be the sum of counts of both old and young carbon as

$$Nc = Na\left(\frac{Ca}{Cm + Ca}\right) + 15.3\left(\frac{Cm}{Cm + Ca}\right)$$

and

$$Nc = \frac{15.3}{Cm + Ca}\left[Ca\, e^{-\frac{0.693ta}{5730}} + Cm\right]$$

From the last equation a date tc is obtained for a contaminated sample. Various values can be substituted for Ca and Cm for samples of known age ta, and a family of curves may be constructed (Fig. 7.1). Contamination of really old carbon by modern carbon will give the most severe effect.

All of these equations, when reduced to curve form,

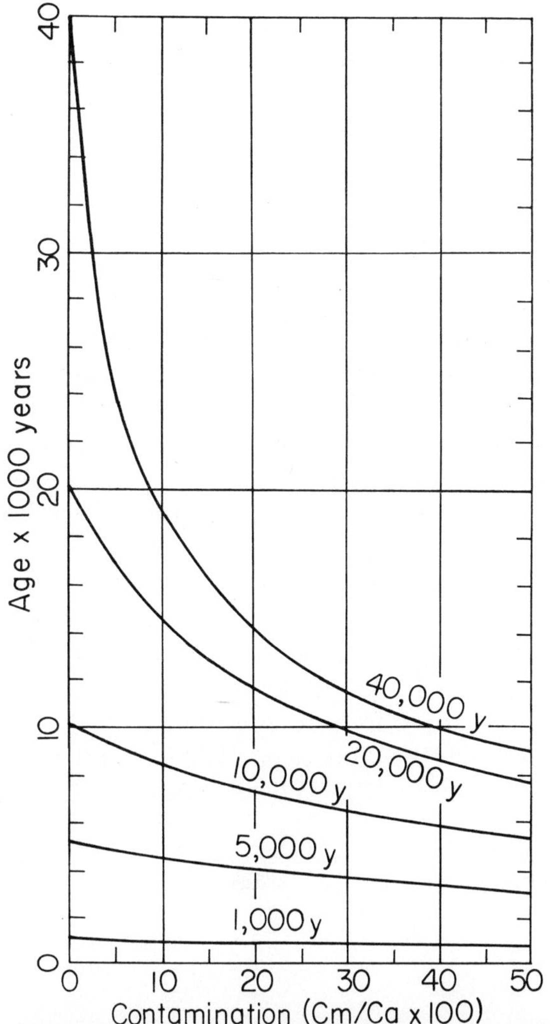

FIG. 7.1. Calculated curves showing the effect on radiocarbon dates by contamination of old carbon by modern carbon. (From Walker, 1965.)

readily show the changes in radiocarbon age that are brought about by contamination. If a sample 40,000 years old has a 5 percent contaminant of modern carbon, the resultant age determined by radiocarbon analysis is 27,000 years. This is a 13,000-year discrepancy. On the other hand, an original sample of 1,000 years may have as much as 50 percent contaminant of modern carbon, and the resultant age is reduced only a few hundred years. Consequently, the older the original sample is, the greater the age discrepancy is with lesser amount of younger contaminant.

These principles are extremely important from the practical point of view of dating of geologic events. For example, a Tazewell feature of 20,000 years ago, if contaminated by 30 percent modern carbon, can date from 10,000 years ago or postglacial time (Fig. 7.1). Not only would the resultant age be obviously erroneous, but it can butcher an otherwise reasonable story.

NATURAL CONTAMINATION

The common procedure in radiocarbon analysis where natural contamination is suspected is to leach the sample with a hot alkaline solution such as NaOH. A date is determined on the more insoluble residue, and a date is determined for more soluble organic constituents that are leached. Comparison is then made between the dates.

Such processing was applied to the organic and mineral sediments of the McCulloch bog (Rubin and Alexander, 1960). Samples are from depths of 6.5 to 7.5, 8.5 to 9.5, and 12.5 to 13 feet. The date of the NaOH-insoluble material of the upper sample is 6,570 ± 200 years (W-551) and the date of the NaOH-soluble material is 6,580 ± 200 years (W-554; see Table 4.3). Comparable values for the middle sample are 8,170 ± 200 (W-549) and 8,110 ± 200 years (W-553) and for the lower sample are 11,660 ± 250 (W-548) and 11,790 ±

250 years (W-552), respectively. The excellent agreement between values of the pairs shows that there has been little if any contamination by younger carbon, and the dates are valid.

Natural contamination by younger organic matter should be expected in A horizons of soils that are buried beneath thin loess, and where a well-developed soil is formed in the upper part of the loess. Such a situation is in Wayne County in southern Iowa. There the Edina soil in loess has a well-formed textural B horizon and has considerable amounts of organic carbon deep in the profile (Fig. 7.2). This organic carbon when

FIG. 7.2. Clay and organic carbon distributions in Edina soil and loess above radiocarbon dated A horizon of basal soil of loess.

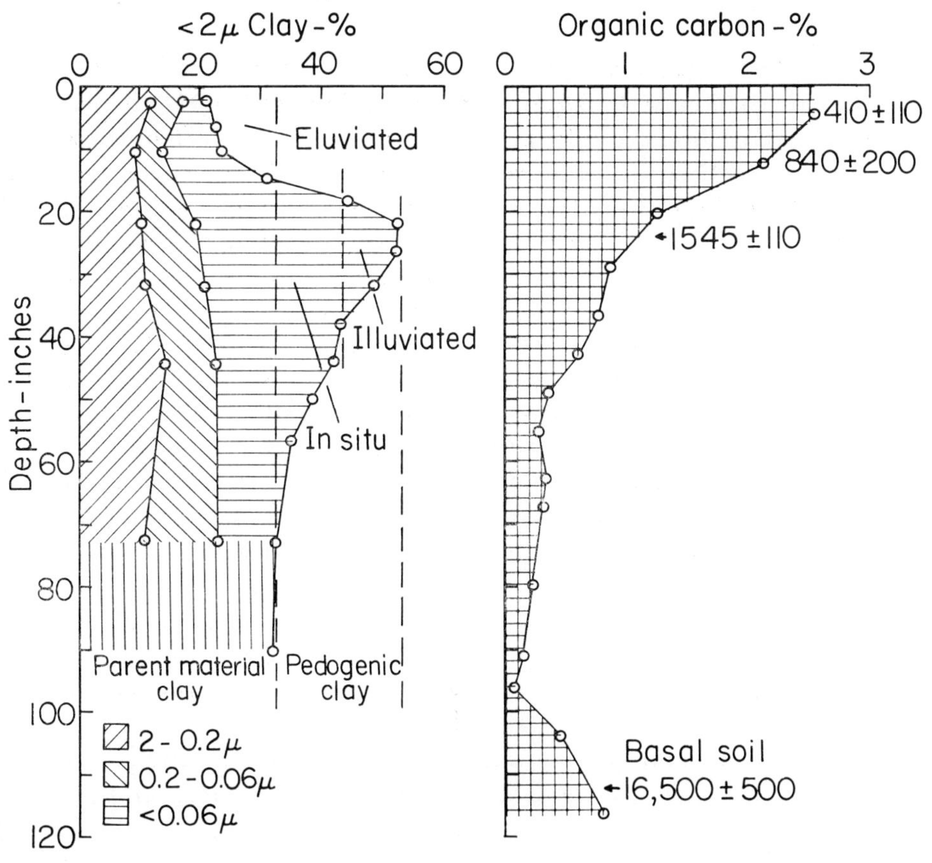

traced downward to 97 inches is as much as 0.1 to 0.2 percent. At this depth the A horizon of the basal soil of the Wisconsin loess is present, and organic carbon increases to 0.6 and 0.8 percent. But, the vertical distribution of the entire organic system suggests the possibility of contamination from above.

The radiocarbon age of the NaOH-insoluble fraction of the organic carbon of the basal soil is 16,500 ± 500 years (I-1419A; see Table 2.3). This date when first available was considerably younger than other dates from the base of the loess that were known at that time. The question of contamination rose again. The value of the NaOH-soluble fraction did not help much, being 19,000 + 6,000 or —3,000 years (I-1419B; see Table 2.3). So, back we go to the first-order equations.

The original sample of 6,000 grams had 0.81 percent organic carbon or 48.6 grams. Leaching by hot NaOH yielded 400 cc of carbon dioxide, or 0.44 percent carbon. When these data are entered into the equation from which our curves are calculated (Fig. 7.1), the maximum possible age of the old carbon *(Ca)* is 17,380 years. This value is based on the assumption that all contaminant was 100 years which it is not. Note the ages of carbon in the A1, A2, and B2 horizons of the surface soil which are 410, 840, and 1,545 years, respectively (Fig. 7.2). The agreement within the whole system shows that little contamination occurs in the basal soil, and the date 16,500 years is valid.

Another case like this involves the dating of the Tazewell till in the Sheldon cut in O'Brien County. The NaOH-insoluble fraction of the organic carbon from the A horizon of a soil beneath calcareous till and thin loess is 20,500 ± 400 years (I-1864A; see Chapter 3). The depth of sample is only 73 inches, so contamination is possible. The NaOH-soluble fraction is 13,400 ± 600 years (I-1864B), indicating that contamination had occurred.

The original sample yielded 10 cc of CO_2 per gram of sample, but the NaOH extract yielded only 0.37 cc of CO_2 per gram of sample. Inserting these data in the equations, the calculated age of the original sample containing both old and young carbon is 20,100 years. This date is practically the same as that of the NaOH-insoluble fraction. Consequently, contamination has not been so severe and the 20,500-year date is valid.

Distinct anomaly is in the sequence of dates from the bogs on the Des Moines lobe (Fig. 5.8). The lower peaty muck and silts generally are 8,000 to 13,000 or more years old. The values in the Hebron bog are 27,900 (I-1853) and 30,300 years (I-1859; see Table 4.3). These dates do not fit and are obviously in error. To determine why, samples of the dated horizons were separated into size grades of 62 to 31μ, 31 to 16μ, and 16 to 2μ and studied under the microscope. Black or carbonaceous shale fragments are common to abundant in all size fractions and comprise 15 to 20 percent of the whole sample in the 31 to 16μ size grade. These shale fragments can be as old as the Cretaceous or Paleozoic, in which the carbon is "dead." It is no wonder, then, that radiocarbon dates of these sediments are so old when they contain such contaminant.

SOIL ORGANIC CARBON

Few published data are available on the variability of radiocarbon ages of organic carbon components in the A horizons of soils. Organic matter is continually added to the A horizon mainly through the decomposition of vegetation that is provided in its cyclic growth. Thus, some organic carbon is older and some is younger. Within the complex organic carbon system in the soil, some parts of it are considered to be "stable" and some are considered to be "relatively mobile and unstable" (Campbell *et al.*, 1967a, b). The various fractions can be separated by chemical extraction with alkaline or acid solutions, and the various fractions can then be radiocarbon dated. See the papers cited for procedures.

Prior to chemical treatment a radiocarbon date of the organic carbon is the "mean residence time of the unfractionated soil" (Campbell *et al.*, 1967a, b). The date of each extracted fraction is the mean residence time (MRT) for that fraction. Comparison can then be made between the ages of extracted fractions and the mean age of the total organic carbon. In one soil in Saskatchewan, Canada (Campbell *et al.*, 1967a, b), where MRT of the unfractionated carbon is 870 ± 50 years, the MRT's of various fractions by various chemical extractions are 25, 325, 465, 470, 495, 555, 785, 870, 1,135, 1,140, 1,230, 1,235, 1,400, and 1,410 ± 45 to 95 years. In another soil where MRT of the unfractionated carbon is 250 ± 60 years, the MRT's of various fractions are 0, 50, 85, 195, 335, and 485 ± 45 to 70 years. Thus, where the MRT is 870 years for unfractionated soil, the range of MRT's of fractions is 25 to 1,410 years and the deviation ranges from 540 to 850 years. Where the MRT is 250 years for unfractionated soil, the range of MRT's of fractions is 0 to 485 years and the deviation ranges from 235 to 250 years.

Now geologically speculating, bury these A horizons under 30 to 40 feet of loess and assume absolute precision in being able to measure the ages 20,000 years from now. The MRT's of unfractionated organic carbon should be 20,250 and 20,870 years. The range in ages of fractions should be 20,000 to 20,485 years for the first soil and 20,025 to 21,410 years for the second soil.

However, the precision of radiocarbon measurement is involved. The half-life of radiocarbon of 5,568 ± 30 years is probably accurate to within 50 years or almost certainly within 100 years (Libby, 1965). Basically there is a 1 to 2 percent error in the immediate accuracy of measurement. A 10,000-year-old sample cannot be measured closer than 100 to 200 years, and a 20,000-year-old sample cannot be measured closer than 200 to 400 years.

Within this framework, the precision of measurement also decreases with the increase in age of sample. The deviations from the age, the ± values, are smaller for younger samples and larger for older samples. In the Iowa dates (Chapter 6) the ranges of deviations group as: (1) 100 to 250 years for samples less than 7,000 years old, (2) 100 to 500 years for samples 7,000 to 13,000 years old, (3) 300 to 1,000 years for samples 13,000 to 21,000 years old, (4) 600 to 1,100 years for samples 21,000 to 25,000 years old, and (5) 1,000 to 3,500 years for samples more than 25,000 years old.

If A horizons of soils with organic carbon properties similar to those in Canada were buried beneath loess more than 14,000 years ago in Iowa, the deviations of ages of organic carbon fractions could not be significantly distinguished from the age of the unfractionated carbon. The deviations inherent in radiocarbon measurement engross the deviations of MRT's of organic carbon fractions from the MRT of the unfractionated carbon. Thus, it can be assumed that the potential error attributable to organic carbon fractions would be negligible. Therefore, the dates of organic carbon of A horizons buried beneath loess in Iowa are reasonable.

CHAPTER 8

RELATIONS

A LARGE PART OF THE PRECEDING discussion has had many numbers inflicted upon it. Many of them have been paired on graph paper—one offhand and the other in the prone position. Some of the pairs are straight and true or a good linear shot. Others require Kentucky windage to hit the target as they are curved. Either way may be the thing necessary to handle some of our relatives.

An analyst in science tries to bore site the relation between paired data by a technique known as *curve fitting*. He may learn to appreciate the exact relation that exists between them or some approximation of it. He may even develop an equation that expresses the relation, but this will require that he recall from the firing line some old acquaintances such as algebra, analytic geometry, and statistics. But out in the country with landscapes and soils, he must appreciate forms in space (Jensen, 1964). Even near a place like River City, Iowa, one has to know the territory before he can sell saxophones.

For most purposes, the best way to fit curves is by the method of *least squares* which requires that the sum of the squares of the error shall be a minimum. This leads to the fitting of a curve called the line of regression of Y on X and leads to the finding of coefficients *a, b* . . . that define the relation.

With this brief background, let us examine some of the paired relations that have been developed previously. Reference can be made to standard texts for further information on background. Our approach will be cookbook style, and a few recipes will be given.

The simplest relation is linear, where one variable of a paired set changes with a constant value versus the other value. An example is the longitudinal profile of the top of the Tazewell alluvium in the Boyer River Valley in southwest Iowa (Fig. 4.16). First, we plot the elevation at a site as Y and the distance of the site in miles above the stream mouth as X. Second, we eyeball the plot, or better yet, lay a transparent straight edge along and through the points. They appear to fall along a straight line. Third, we set up a work chart to find the fit of the line through the points and the empirical equation of the line (Table 8.1). The values needed are ΣX, ΣY,

TABLE 8.1. CALCULATION OF LONGITUDINAL PROFILE OF TOPS OF TAZEWELL ALLUVIUM IN BOYER RIVER VALLEY, IOWA, AND ITS EMPIRICAL EQUATION

X	Y	XY	X^2	Y^2	Yc	$Y - Yc$	$(Y - Yc)^2$
10.3	994				980.8	13.2	
13.2	1004				997.0	7.0	
16.1	1011				1013.2	—2.2	
18.9	1031				1028.9	2.2	
22.3	1040				1047.9	—7.8	
23.2	1046				1052.9	—6.8	
26.5	1072				1071.3	0.7	
31.6	1098				1099.8	—1.7	
32.3	1090				1103.7	—13.6	
32.5	1095				1104.8	—9.8	
35.1	1123				1119.3	3.7	
40.1	1138				1147.3	—9.2	
40.4	1151				1148.9	2.1	
41.4	1159				1154.5	4.5	
44.1	1182				1169.6	12.4	
45.6	1180				1177.9	2.1	
50.0	1197				1202.5	—5.5	
54.0	1232				1224.9	7.2	
54.9	1231				1229.9	1.1	
55.4	1234				1232.7	1.3	
687.9	22,308	788,404	27,442	25,001,292			990.53

Where X is distance of site in miles above stream mouth.
Y is elevation of top of Tazewell alluvium in feet above sea level.
Yc is calculated value of Y from empirical equation.

ΣXY, ΣX^2, ΣY^2, and N, the last being the number of paired measurements. These values, excluding ΣY^2, are inserted into the normal equations of the least squares system

$$\Sigma Y = aN + b\Sigma X$$
$$\Sigma XY = a\Sigma X + b\Sigma X^2$$

which are solved simultaneously, giving for our problem

$$Y = 923.3 + 5.59X$$

which means that from a site at an elevation of 923.3 feet above sea level, the longitudinal profile slopes upstream at the rate of 5.59 feet per mile.

The spread of the measured points from the calculated fitted curve may be determined by calculating values of Y *(Yc)* by inserting the measured values of X and solving. The differences are then $Y - Yc$. For the whole set of measurements a *standard error of estimate (Sy)* is calculated as

$$Sy = \sqrt{\frac{\Sigma(Y - Yc)^2}{N}}$$

In our problem, this value is about 7 feet. This means that the empirical equation estimates that about two-thirds of the measured values will be within 7 feet above or below the predicted longitudinal profile. This is not bad considering that a distance of almost 50 miles is involved along the valley.

Other statistical measures can be made such as coefficient of correlation *(r)* of the data, and ΣY^2 enters this calculation. In our problem this value is 0.996. If exact fit existed, the value would be 1.0, but complications enter here and reference should be made to standard texts in statistics.

Commonly, relations are not straight line but curved, so there are crooked relations and skeletons in the closet. One such case is the thinning of the loess with distance from the Missouri River Valley in southwest Iowa (Fig. 4.2A). The curved relation is seen when loess thickness Y is plotted versus

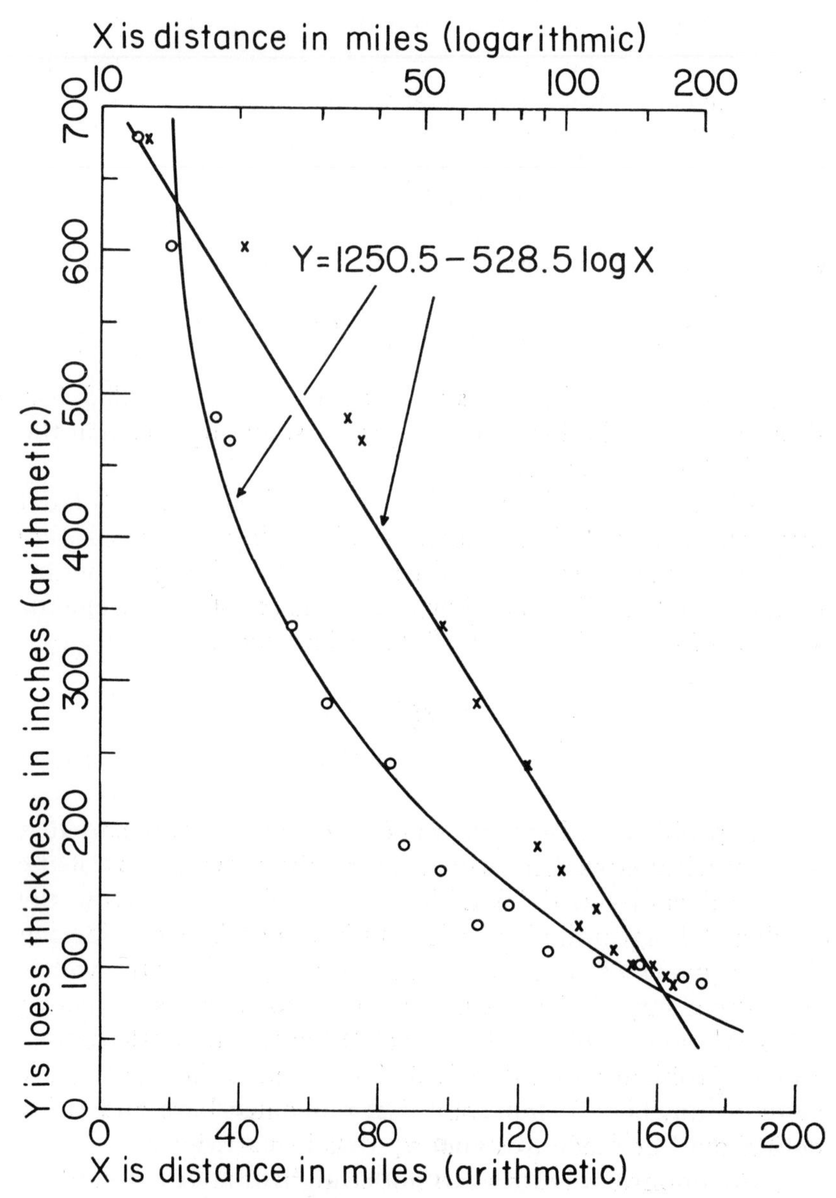

FIG. 8.1. Loess thickness decreases with distance from the source in southwest Iowa. Plotted on linear scales, the relation is a curve. Plotted on semilogarithmic scale, the relation appears as a straight line.

distance X with both axes in linear units (Fig. 8.1). Reference can be made at this point to charts of curves and their kinds of empirical equations. It also helps if one recalls that a logarithm is the exponent that indicates the power to which a number is raised to produce a given number. With these aids, the loess thickness and distance data are replotted with Y in linear units and X as the logarithm of distance. The eyeballed points on the semilog paper plot seem to fall along a transparent straight edge. A line can be fitted and tested by going through the recipe of the cookbook (Table 8.1). However, log X is used as an ingredient rather than X. After all is done, the calculated line is expressed by

$$Y = 1250.5 - 528.5 \log X$$

If desired, one may go on and test the goodness of fit statistically.

An estimate may be made of other kinds of curves and empirical equations by using other kinds of or other orientation of axes on graph paper. If Y varies exponentially with linear variation of X, the values of Y are plotted on the logarithmic scale and values of X are plotted on the linear scale of semilogarithmic paper. Both variables may change exponentially so plots may be made on logarithmic paper where each scale is logarithmic. Commonly, a curve may be hyperbolic, such as the thinning of loess along the Rock Island Railroad cuts from Bentley to Adair, Iowa. A fitted curve is defined by the reciprocal of Y or $1/Y$ versus a linear scale of X. Where both scales are linear, the relation plots as a curve (Fig. 2.3). However, if reciprocal ruled graph paper is used and the axes are properly oriented, the relation plots as a straight line. To calculate this curve, one need only substitute values of $1/Y$ for Y in the charted recipe and have at it.

In some cases ordinary empirical equations such as the linear, exponential, power, and polynomial kinds do not explain the relation. One must resort to scaling and the direct determination of algebraic form components. The fits of soils to hillslopes in the open system are cases in point (Fig. 4.9). Following the techniques of Jensen (1964), we can examine the fit of clay zone thickness to slope gradient in Adair County (Fig. 4.8). First, the paired data are plotted on linear scale graph paper (Fig. 8.2). In order to avoid negative values, the X scale is transposed to $(10 - X)$. By least squares, two linear relations can be fitted to the data

$$A \text{ is } Y = 21.63 + 6.64\ (10 - X)$$
$$B \text{ is } Y = 20.65 + 1.23\ (10 - X)$$

The two linear relations join at a point marked by slope gradient of about 2 or $10 - X = 8$ percent and clay zone thickness of 30 inches. This bipartite fitting may be satisfactory for some purposes, but an overall fit of one curve to all data may be desirable. So, we must set up a work sheet (Table 8.2) and try to bend a curve through the two linear subsets. More specifically, we need only bend a curve through subset A that joins subset B. Curvature is brought about by adding terms with higher powers of $(10 - X)$ as

$$Y = 20.65 + 1.23\ (10 - X) + b_2\ (10 - X)^n$$

To add the next term, Y must be described in some term of $(10 - X)$ by scaling a given value of $(10 - X)$ to Y.

With subset B determined only, $Y - Y_1$ is 13.05 where $(10 - X)$ equals 10 (Table 8.2); Y_1 is the calculated value of Y from the linear equation. A curve must minimize this difference. So, we will try $(10 - X)^4$ which for value 10 is 10,000. Thus,

$$\frac{13.05}{10{,}000} \times 10{,}000 = 13.05 \text{ or } Y_2$$

and $Y_1 + Y_2 = 32.95 + 13.05 = 46.00$ or $Y - (Y_1 + Y_2) = 46 - 46 = 0$. This scaling of $(10 - X)$ or 10 for the Y value of 46 is the ultimate minimum, and the fit is perfect. But, for the other pairs in the set the $Y - (Y_1 + Y_2)$ values are not so good (Table 8.2) and the curve does not fit (Fig. 8.2). The average difference is 2.3 inches on Y. So, $(10 - X)^4$ is not a reasonable third term in the equation.

Next, we try $(10 - X)^8$ to get more curvature and follow

FIG. 8.2. Curve fitting by scaling and the direct determination of algebraic form components. (Cf. Fig. 4.9.)

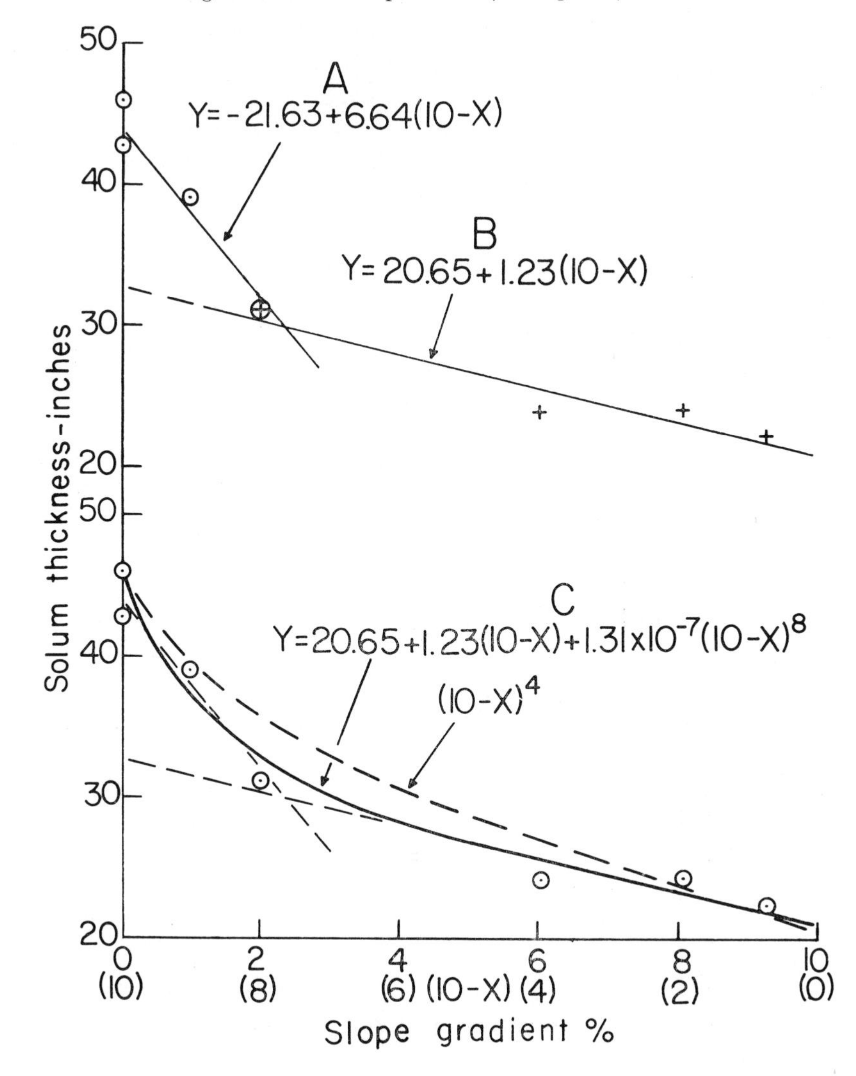

TABLE 8.2. WORK SHEET FOR SCALING AND DIRECT DETERMINATION OF ALGEBRAIC FORM COMPONENTS TO RELATE VARIABLES

X	$(10 - X)$	Y	Y_1 $20.65 + 1.23\ (10 - X)$	$Y - Y_1$	Try $(10-X)^4$	Y_2 $(13.05/10^4)\ (10-X)^4$ $(1.31 \times 10^{-3})\ (10-X)^4$	$Y_1 + Y_2$	$Y - (Y_1 + Y_2)$
0	10	46	32.95	13.05	10,000	13.05	46.00	0.00
0	10	43	32.95	10.05	10,000	13.05	46.00	—3.00
1	9	39	31.72	7.28	6,561	8.56	40.28	—1.28
2	8	31	30.49	0.51	4,096	5.35	35.84	—4.84
6	4	24	25.57	—1.57	1,296	1.69	27.26	—3.26
8	2	24	23.11	0.89	16	0.02	23.13	0.87
9	1	22	21.88	0.12	1	0.00	21.88	0.12
					Try $(10 - X)^8$	Y_2 $(13.05/10^8)\ (10 - X)^8$ $(1.31 \times 10^{-7})\ (10 - X)^8$		
					100,000,000	13.05	46.00	0.00
					100,000,000	13.05	46.00	—3.00
					43,046,721	5.61	37.33	1.67
					16,777,216	2.19	32.68	—1.68
					65,536	0.01	25.58	—1.58
					256	0.00	23.11	0.89
					1	0.00	21.88	0.12

Where X is slope gradient in percent and Y is solum thickness in inches.

a similar calculating procedure for all pairs of the data. The curve fits better (Fig. 8.2C) and the differences in $Y - (Y_1 + Y_2)$ have been further minimized (Table 8.2). The average difference is 1.5 inches on Y. The third term to be added to the equation is $+ 1.31 \times 10^{-7} (10 - X)^8$ and our data are organized.

So with all of these manipulations, we have straightened out our relations and that is frequently needed in a family. Have a try at the cookbook and see what kind of table can be set.

REFERENCES

ALDEN, W. C., and LEIGHTON, M. M., "The Iowan Drift, a Review of the Evidences of the Iowan Stage of Glaciation." *Iowa Geol. Survey Ann. Rept.* 26:49–212, 1917.

AMER. COMM. STRAT. NOMEN., "Code of Stratigraphic Nomenclature." *Amer. Assoc. Petrol. Geol. Bull.* 45:645–55, 1961.

ANDERSSON, J. G., "Solifluction, a Component of Subaerial Denudation." *Jour. Geology* 14:91–112, 1906.

BRAY, R. H., "A Chemical Study of Soil Development in the Peorian Loess Region in Illinois." *Amer. Soil Survey Assoc. Bull.* 15:58–65, 1934.

———, "The Origin of Horizons in Claypan Soils." *Amer. Soil Survey Assoc. Bull.* 16:75–85, 1935.

BRUSH, G. S., "Pollen Analyses of Late-Glacial and Postglacial Sediments in Iowa." *Quaternary Paleoecology,* Yale Univ. Press, 99–115, 1967.

BRYAN, KIRK, "Cryopedology—the Study of Frozen Ground and Intensive Frost Action With Suggestions on Nomenclature." *Amer. Jour. Sci.* 244:622–42, 1946.

CALVIN, SAMUEL, "Iowan Drift." *Geol. Soc. Amer. Bull.* 10:107–20, 1899.

CAMPBELL, C. A., PAUL, E. A., RENNIE, D. A., and MCCULLUM, K. J., "Applicability of the Carbon-dating Method of Analysis to Soil Humus Studies." *Soil Sci.* 104:217–24, 1967a.

———, "Factors Affecting the Accuracy of the Carbon-dating Method in Soil Humus Studies." *Soil Sci.* 104:81–85, 1967b.

CARMAN, J. E., "Further Studies on the Pleistocene Geology of Northwestern Iowa." *Iowa Geol. Survey Ann. Rept.* 35:15–193, 1931.

CARMAN, J. E., "The Pleistocene Geology of Northwestern Iowa." *Iowa Geol. Survey Ann. Rept.* 26:233–445, 1917.

CHAMBERLIN, T. C., "The Classification of American Glacial Deposits." *Jour. Geology* 3:270–77, 1895.

CLINE, M. G., "Basic Principles of Soil Classification." *Soil Sci.* 67:81–91, 1949.

CORLISS, J. F., and RUHE, R. V., "The Iowan Terrace and Terrace Soils of the Nishnabotna Valley in Western Iowa." *Iowa Acad. Sci. Proc.* 62:345–60, 1955.

CUSHING, E. J., "Problems in the Quaternary Phytogeography of the Great Lakes Region." *The Quaternary of the United States,* INQUA rev. vol., Princeton Univ. Press, 403–16, 1965.

DANIELS, R. B., and JORDAN, R. H., "Physiographic History and the Soils, Entrenched Stream Systems, and Gullies, Harrison County, Iowa." *U.S. Dept. Agric. Tech. Bull.* 1348, 1966.

DANIELS, R. B., SIMONSON, G. H., and HANDY, R. L., "Ferrous Iron Content and Color of Sediments." *Soil Sci.* 91:378–82, 1961.

DAVIS, M. B., "On the Theory of Pollen Analysis." *Amer. Jour. Sci.* 261:897–912, 1963.

———, "Phytogeography and Palynology of Northeastern United States." *The Quaternary of the United States,* INQUA rev. vol., Princeton Univ. Press, 377–401, 1965.

EMILIANI, CESARE, "Paleotemperature Analysis of Caribbean Cores P6304-8 and P6304-9 and a Generalized Temperature Curve for the Past 425,000 Years." *Jour. Geology* 74:109–24, 1966.

FLINT, R. F., *Glacial and Pleistocene Geology.* John Wiley & Sons, Inc., New York, 1957.

———, "Pleistocene Geology of Eastern South Dakota." *U.S. Geol. Survey Prof. Paper* 262, 1955.

FLINT, R. F., and RUBIN, MEYER, "Radiocarbon Dates of Pre-Mankato Events in Eastern and Central North America." *Science* 121:649–58, 1955.

FRYE, J. C., and LEONARD, A. B., "Pleistocene Stratigraphic Sequence in Northeastern Kansas." *Amer. Jour. Sci.* 247:883–99, 1949.

FRYE, J. C., and LEONARD, A. R., "Some Problems of Alluvial Terrace Mapping." *Amer. Jour. Sci.* 252:242–51, 1954.

FRYE, J. C., and WILLIAM, H. B., "Classification of the Wisconsinan Stage in the Lake Michigan Glacial Lobe." *Illinois Geol. Survey Circ.* 285, 1960.

GWYNNE, C. S., "Minor Moraines in South Dakota and Minnesota." *Geol. Soc. Amer. Bull.* 62:233–50, 1951.

———, "Swell and Swale Pattern of the Wisconsin Drift Plain in Iowa." *Jour. Geology* 50:200–208, 1942.

GWYNNE, C. S., and SIMONSON, R. W., "Influence of Low Recessional Moraines on Soil Type Pattern of the Mankato Drift Plain in Iowa." *Soil Sci.* 53:461–66, 1942.

HANNA, R. M., and BIDWELL, O. W., "The Relation of Certain Loessial Soils of Northeastern Kansas to the Texture of the Underlying Loess." *Soil Sci. Soc. Amer. Proc.* 19:354–59, 1955.

HEUSSER, C. J., "A Pleistocene Phytogeographical Sketch of the Pacific Northwest and Alaska." *The Quaternary of the United States,* INQUA rev. vol., Princeton Univ. Press, 469–83, 1965.

HORTON, R. E., "Erosional Development of Streams and Their Drainage Basins–Hydrophysical Approach to Quantitative Morphology." *Geol. Soc. Amer. Bull.* 56:275–370, 1945.

HOVDE, M. R., "Climate of Minnesota." *Climate and Man,* U.S. Dept. Agric. Yearbook of Agriculture, 925–34, 1941.

HOWARD, A. D., "Numerical Systems of Terrace Nomenclature, a Critique." *Jour. Geology* 67:239–43, 1959.

HUTTON, C. E., "Studies of the Chemical and Physical Characteristics of a Chrono-Litho-Sequence of Loess-Derived Prairie Soils of Southwestern Iowa." *Soil Sci. Soc. Amer. Proc.* 15:318–24, 1951.

———, "Studies of Loess-Derived Soils in Southwestern Iowa." *Soil Sci. Soc. Amer. Proc.* 12:424–31, 1947.

JENNY, HANS, *Factors of Soil Formation.* McGraw-Hill Co., Inc., New York, 1941.

JENSEN, C. E., "Algebraic Description of Forms in Space." U.S. Dept. Agric., Forest Serv., *Central States Exp. Sta. Circ.,* 1964.

JOHNSON, D. W., "Problems of Terrace Correlation." *Geol. Soc. Amer. Bull.* 55:793–818, 1944.

KAY, G. F., "Classification and Duration of the Pleistocene Period." *Geol. Soc. Amer. Bull.* 42:425–66, 1931.

———, "Gumbotil, a New Term in Pleistocene Geology." *Science,* new ser. 44:637–38, 1916.

———, "Pleistocene Deposits Between Manilla in Crawford County and Coon Rapids in Carroll County, Iowa." *Iowa Geol. Survey Ann. Rept.* 26:213–31, 1917.

KAY, G. F., and APFEL, E. T., "The Pre-Illinoian Pleistocene Geology of Iowa." *Iowa Geol. Survey Ann. Rept.* 34:1–304, 1929.

KAY, G. F., and GRAHAM, J. B., "The Illinoian and Post-Illinoian Pleistocene Geology of Iowa." *Iowa Geol. Survey Ann. Rept.* 38:1–262, 1943.

KAY, G. F., and MILLER, P. T., "The Pleistocene Gravels of Iowa." *Iowa Geol. Survey Ann. Rept.* 37:1–231, 1941.

KAY, G. F., and PEARCE, J. N., "The Origin of Gumbotil." *Jour. Geology* 28:89–125, 1920.

KOHMAN, T. P., and SAITO, NOBUFUSA, "Radioactivity in Geology and Cosmology." *Ann. Rev. Nuclear Sci.* 4:401–62, 1954.

KULP, J. L., "Geologic Time Scale." *Science* 133:1105–14, 1961.

LAWRENCE, D. B., and ELSON, J. A., "Periodicity of Deglaciation in North America Since the Late Wisconsin Maximum." *Geografiska Ann.* 35:83–104, 1953.

LEES, J. H., "Physical Features and Geologic History of Des Moines Valley." *Iowa Geol. Survey Ann. Rept.* 25:423–615, 1916.

LEIGHTON, M. M., "The Classification of the Wisconsin Glacial Stage of North Central United States." *Jour. Geology* 68:529–52, 1960.

———, "Important Elements in the Classification of the Wisconsin Glacial Stage." *Jour. Geology* 66:288–309, 1958.

———, "The Naming of the Subdivisions of the Wisconsin Glacial Age." *Science,* new ser. 77:168, 1933.

———, "The Peorian Loess and the Classification of the Glacial Drift Sheets of the Mississippi Valley." *Jour. Geology* 39:45–53, 1931.

———, "Review of Papers on Continental Glaciation, INQUA Volume on the Quaternary." *Jour. Geology* 74:939–46, 1966.

LEIGHTON, M. M., and WILLMAN, H. B., "Loess Formations of the Mississippi Valley." *Jour. Geology* 58:599–623, 1950.

LEONARD, A. B., "Illinoian and Wisconsinan Molluscan Faunas." *Univ. Kansas Paleontol. Contrib.* Art. 4, 1952.

LEOPOLD, L. B., WOLMAN, M. G., and MILLER, J. P., *Fluvial Processes in Geomorphology.* W. H. Freeman and Co., San Francisco, 1964.

LEVERETT, FRANK, "Editorial." *Jour. Geology* 4:872–76, 1896.

———, "Note by Frank Leverett." *Jour. Geology* 50:1001–2, 1942.

———, "The Place of the Iowan Drift." *Jour. Geology* 47:398–407, 1939.

———, "The Pleistocene Glacial Stages; Were There More Than Four?" *Amer. Philos. Soc. Proc.* 65:105–18, 1926.

———, "The Weathered Zone (Yarmouth) Between the Illinoian and Kansan Till Sheets." *Jour. Geology* 6:238–43, 1898.

———, "Weathering and Erosion As Time Measures." *Amer. Jour. Sci.* 4th ser. 27:349–68, 1909.

LIBBY, W. F., "Radiocarbon Dating." *Science* 133:621–29, 1960.

———, *Radiocarbon Dating.* Univ. Chicago Press, Chicago, 2d ed., 1965.

MARBUT, C. F., *Soils: Their Genesis and Classification.* Soil Sci. Soc. Amer., Madison, Wisc., 1951.

MARTIN, P. S., and MEHRINGER, P. J., "Pleistocene Pollen Analysis and Biogeography of the Southwest." *The Quaternary of the United States,* INQUA rev. vol., Princeton Univ. Press, 433–51, 1965.

McGee, WJ, "The Pleistocene History of Northeastern Iowa." *U.S. Geol. Survey 11th Ann. Rept.* :189–577, 1891.

Olson, E. A., and Broecker, W. S., "Sample Contamination and Reliability of Radiocarbon Dates." *New York Acad. Sci. Trans.* 20:593–604, 1958.

Oschwald, W. R., Riecken, F. F., Dideriksen, R. I., Scholtes, W. H., and Schaller, F. W., "Principal Soils of Iowa." *Iowa State Univ. Ext. Serv. Spec. Rept.* 42, 1965.

Péwé, T. L., "Notes on the Physical Environment of Alaska." *Proc. 15th Alaska Sci. Conf.* :293–310, 1965.

Reed, C. D., "Climate of Iowa." *Climate and Man,* U.S. Dept. Agric. Yearbook of Agriculture, 862–72, 1941.

Richmond, G. M., and Frye, J. C., "Status of Soils in Stratigraphic Nomenclature." *Amer. Assoc. Petrol. Geol. Bull.* 41:758–62, 1957.

Robinson, G. W., *Soils, Their Origin, Constitution, and Classification.* Thomas Murby Co., London, 3d ed., 1949.

Rubin, Meyer, and Alexander, Corrine, "U.S. Geological Survey Radiocarbon Dates V." *Amer. Jour. Sci. Radiocarb. Suppl.* 2:129–85, 1960.

Ruhe, R. V., "Classification of the Wisconsin Glacial Stage." *Jour. Geology* 60:398–401, 1952b.

———, "Geomorphic Surfaces and the Nature of Soils." *Soil Sci.* 82:441–55, 1956.

———, "Geomorphic Surfaces and Surficial Deposits in Southern New Mexico." *New Mexico Bur. Mines & Min. Res.* Mem. 18, 1967.

———, "Graphic Analysis of Drift Topographies." *Amer. Jour. Sci.* 248:435–43, 1950.

———, "The Iowa Quaternary." *INQUA Guidebook for Field Conf. C, Upper Mississippi Valley* :110–26, 1965b.

———, "Quaternary Paleopedology." *The Quaternary of the United States,* INQUA rev. vol., Princeton Univ. Press, 755–64, 1965a.

———, "Relations of the Properties of Wisconsin Loess to Topography in Western Iowa." *Amer. Jour. Sci.* 252:663–72, 1954.

———, "Stone Lines in Soils." *Soil Sci.* 87:223–31, 1959.

———, "Topographic Discontinuities of the Des Moines Lobe." *Amer. Jour. Sci.* 250:46–56, 1952a.

Ruhe, R. V., and Daniels, R. B., "Landscape Erosion—Geologic and Historic." *Jour. Soil & Water Conserv.* 20:52–57, 1965.

Ruhe, R. V., and Scholtes, W. H., "Ages and Development of Soil Landscapes in Relation to Climatic and Vegetational Changes in Iowa." *Soil Sci. Soc. Amer. Proc.* 20:264–73, 1956.

RUHE, R. V., and SCHOLTES, W. H., "Important Elements in the Classification of the Wisconsin Glacial Stage—A Discussion." *Jour. Geology* 67:585–93, 1959.

RUHE, R. V., DANIELS, R. B., and CADY, J. G., "Landscape Evolution and Soil Formation in Southwestern Iowa." *U.S. Dept. Agric. Tech. Bull.* 1349, 1967.

RUHE, R. V., RUBIN, MEYER, and SCHOLTES, W. H., "Late Pleistocene Radiocarbon Chronology in Iowa." *Amer. Jour. Sci.* 255:671–89, 1957.

RUHE, R. V., DIETZ, W. P., FENTON, T. E., and HALL, G. F., "Iowan Drift Problem, Northeastern Iowa." *Iowa Geol. Survey Rept. Inv.* 7, 1968.

SAVIGEAR, R. A. G., "Technique and Terminology in the Investigation of Slope Forms." *Int. Geog. Union Comm. Study of Slopes 1st Rept.* :66–75, 1956.

SCHAFER, G. M., "A Disturbed Buried Gumbotil Soil Profile in Jefferson County, Iowa." *Iowa Acad. Sci. Proc.* 60:403–7, 1953.

SCHOEWE, W. H., "The Origin and History of Extinct Lake Calvin." *Iowa Geol. Survey Ann. Rept.* 29:49–222, 1920.

SCHOLTES, W. H., "Properties and Classification of the Paha Loess-Derived Soils in Northeastern Iowa." *Iowa State Univ. Jour. Sci.* 30:163–209, 1955.

SHAFFER, P. R., "Extension of Tazewell Glacial Substage of Western Illinois and Eastern Iowa." *Geol. Soc. Amer. Bull.* 65:443–56, 1954.

SHIMEK, BOHUMIL, "Aftonian Sands and Gravels in Western Iowa." *Geol. Soc. Amer. Bull.* 20:339–408, 1909.

———, "Geology of Harrison and Monona Counties, Iowa." *Iowa Geol. Survey Ann. Rept.* 20:271–485, 1910.

SIMONSON, R. W., "Identification and Interpretation of Buried Soils." *Amer. Jour. Sci.* 252:705–32, 1954.

———, "What Soils Are." *Soils,* U.S. Dept. Agric. Yearbook of Agriculture :17–31, 1957.

SIMONSON, R. W., and GARDINER, D. R., "Concept and Functions of the Pedon." *Trans. 7th Int. Cong. Soil Sci.* 4:127–31, 1960.

SMITH, G. D., "Illinois Loess, Variations in Its Properties and Distribution—A Pedologic Interpretation." *Illinois Agric. Exp. Sta. Bull.* 490:139–84, 1942.

SMITH, G. D., and RIECKEN, F. F., "The Iowan Drift Border of Northwestern Iowa." *Amer. Jour. Sci.* 245:706–13, 1947.

SOIL SURVEY STAFF, *Soil Survey Manual.* U.S. Dept. Agric. Handbook 18, 1951.

STAFF, *Soils and Men.* U.S. Dept. Agric. Yearbook of Agriculture, 1938.

STRAHLER, A. N., "Quantitative Slope Analysis." *Geol. Soc. Amer. Bull.* 62:571–96, 1956.

TATOR, B. A., "Pediment Characteristics and Terminology." *Assoc. Amer. Geog. Ann.* 43:47–53, 1953.

THORNBURY, W. D., *Principles of Geomorphology.* John Wiley & Sons, Inc., New York, 1954.

THORP, JAMES, and SMITH, H. T. U., "Pleistocene Eolian Deposits of the United States, Alaska, and Parts of Canada." *Geol. Soc. Amer. Map,* 1952.

TROLL, CARL, "Structure Soils, Solifluction, and Frost Climates of the Earth." *U.S. Army Corps Engrs. Snow, Ice, and Permafrost Res. Estab. Transl.* 43 (by H. E. Wright), 1958.

TROWBRIDGE, A. C., "The Erosional History of the Driftless Area." *Univ. Iowa Studies Nat. History* 9:1–127, 1921.

ULRICH, RUDOLPH, "Some Chemical Changes Accompanying Profile Formation of the Nearly Level Soils Developed From Peorian Loess in Southwestern Iowa." *Soil Sci. Soc. Amer. Proc.* 15:324–29, 1951.

———, "Some Physical Changes Accompanying Prairie, Wiesenboden, and Planosol Soil Profile Development From Peorian Loess in Southwestern Iowa." *Soil Sci. Soc. Amer. Proc.* 14:287–95, 1950.

WALKER, P. H., "Postglacial Environments in Relation to Landscape and Soils on the Cary Drift, Iowa." *Iowa Agric. Exp. Sta. Res. Bull.* 549:838–75, 1966.

WALKER, P. H., and BRUSH, G. S., "Observations on Bog and Pollen Stratigraphy of the Des Moines Glacial Lobe, Iowa." *Iowa Acad. Sci. Proc.* 70:253–60, 1963.

WASCHER, H. L., HUMBERT, R. P., and CADY, J. G., "Loess in the Southern Mississippi Valley—Identification and Distribution of the Loess Sheets." *Soil Sci. Soc. Amer. Proc.* 12:389–99, 1947.

WASHBURN, A. L., "Classification of Patterned Ground and Review of Suggested Origins." *Geol. Soc. Amer. Bull.* 67:823–66, 1956.

WATTS, W. A., and WINTER, T. C., "Plant Macrofossils From Kirchner Marsh, Minnesota—A Paleoecological Study." *Geol. Soc. Amer. Bull.* 77:1339–60, 1966.

WHITEHEAD, D. R., "Palynology and Pleistocene Phytogeography of Unglaciated Eastern North America." *The Quaternary of the United States,* INQUA rev. vol., Princeton Univ. Press, 417–32, 1965.

WILLS, H. M., "Climate of Michigan." *Climate and Man,* U.S. Dept. Agric. Yearbook of Agriculture :914–24, 1941.

AUTHOR INDEX

SUBJECT INDEX